Liquid Solid Separation with Filter Presses

Copyright © 2006 Hans G. Valerius PhD

All rights reserved. No part of this publication may be reproduced or transmitted in any form or by any means, electronic or mechanical, including photocopy, recording, or any information storage and retrieval system, without permission in writing from the publisher or under the license from the Copyright Licensing Agency Limited.
ISBN: 1-4196-3395-3

To order additional copies, please contact us.
BookSurge, LLC
www.booksurge.com
1-866-308-6235
orders@booksurge.com

HANS G. VALERIUS
PHD

LIQUID SOLID SEPARATION WITH FILTER PRESSES

Liquid Solid Separation with Filter Presses and Filter Elements

2006

Liquid Solid Separation with Filter Presses

CONTENTS

Introduction	xiii
Liquid/solid Separation Processes	1
Application Of Filter Presses	5
Function Of Filter Presses	7
Construction And Design Of Filter Presses	9
Processing In A Filter Press	15
Optimization Of A Filter Press Cycle	17
Minimizing Of Down Time	19
Discharge Time	21
Mobile Filter Press	23
Filter Press Accessories	25
Filter Cloth Washing	27
Filter Cloth Washing Devices	29
Automatic Cake Discharge	31
Cake Removal Devices	33
Safety Devices On Filter Presses	35
Size And Capacity Of Filter Presses	37
How To Size A Pump	39
Application Of Filter Press—general	41
General Slurry Tests	43
Necessary Data For Sizing The Filter Press	47
Necessary Test Filtration	49
Trouble Shooting	53
Maintenance In General	57
Flow Charts	APPENDIX I
Flow Charts Corner Feed	APPENDIX II
Flow Charts Center Feed	APPENDIX III

FILTER ELEMENTS

Filter Elements For Discontinuous Separation Systems	59
Filter Elements For Filter Presses	61
Filter Elements With Fixed Chamber Depth	63
Chamber plates	
Caulked and gasketed filter elements	65
Filter Elements With Variable Chamber Depth	65
Comparison	66
Membrane Plates	67
Design Of The Membrane Plate	68
Mixed Pack	70
Differential Pressures	70
Material For Filter Elements	71
Material For Membrane Plates	72
Effects On A Complete Filter Press	77
Comparison Of Different Materials For Filter Elements	77
Production Of The Filter Elements	79
Application Of The Different Filter Element Types	80
Cake Washing	82
Monetary Aspects	86
Non—monetary Aspects	88
Squeeze System	89
Membrane Air Manifold	92
Air Blowing	95
Core Wash / Blow	96
Venting	96
Line Blow	96
Sterilization Of Membrane Plates	96
Accessories	96
Technical Clarification	97
Operating Filter Elements	98
Trouble Shooting	103

Out of balance pressure	106
DIN standard	108
Physical Properties Of Hostalen Pp Grades	119
Material Properties Nbr Rubber	123
Squeeze Pressure Membrane Filter Plates	124
Filtration And Wash Pressure Membrane Filter Plates	125
Calculation Of Slurry Throughput	129
Criteria For Membrane Selection	132
Example Filter Presses For Treatment Of Sludges	146
Filter Presses	147
Sludge Conditioning	150

INTRODUCTION

The following description is divided into three major categories, the application of filter presses, details on filter elements and example of application of filter presses in sludge treatment.

The content should contribute to a better understanding of the systems involved and should provide engineers and operators sufficient details to have a proper basis prior to any investment decision and to succeed in a proper operation of the system. After nearly 150 years of existence of filter presses the available options are so ample that for an outsider it will always be difficult to understand the pros and cons of the individual possibilities given. Today filter presses can be automated to a high degree while at the same time other filtration and separation devices also experience further development and might even offer advantages over filter presses in terms of operation and maintenance costs as well as filtration/separation results.

In combination with the right choice of filter elements high dry solid contents and noticeable reduction in filtration time can be achieved.

The liquid-solid separation with filter elements that are the most important part in the filter press comprises a variety of process steps that are more or less uniform, however a complicated matter when it comes to selection of filter element type, cake thickness, infeed system and material selection. The separation process of liquids and solids for one product is not necessarily duplicated for another product of the same structure, i.e. the parameters for one product in comparison with another product of the same type can be totally different and even within one product, there might be alterations required in the process, when the product has a different particle size. Influences of pressure and temperature as well as feeding system are additional parameters to choose from.

Since there is no real basis for a method of calculation for the filtration output, results etc. tests should always be conducted to find the best suitable process.

In spite of these small barriers, liquid-solid separation is neither a science nor the result of any magic work.

The indicated results in the following content are based on practical tests and are subject to product specification, location and equipment used. They are to be considered as guidelines only.

The breakdown of filtration/separation devices refer to various possibilities but mainly within the range of filter presses a comparison can be done subject to the filtration/separation task with belt presses, decanters and screw presses. In some cases also with so-called stainless steel sheet filters using filter paper as filter media. It very much depends on the product to be handled.

LIQUID/SOLID SEPARATION PROCESSES

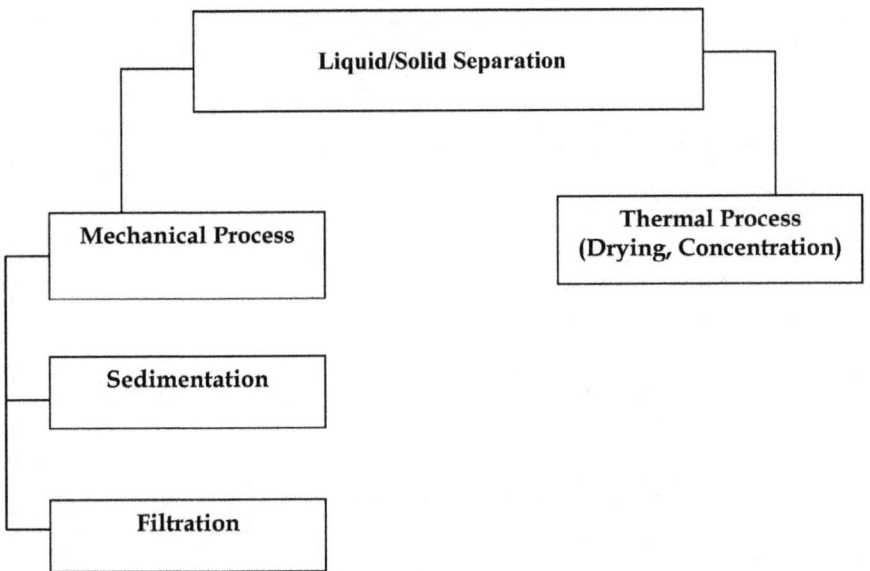

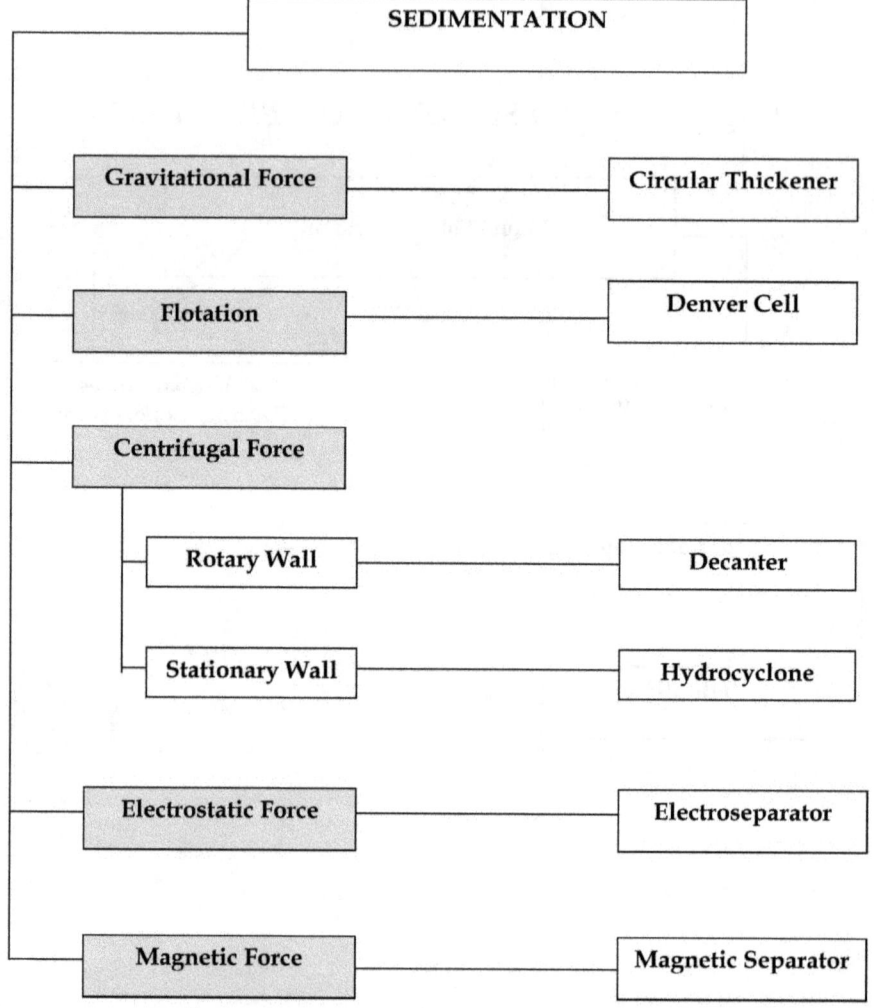

LIQUID SOLID SEPARATION WITH FILTER PRESSES

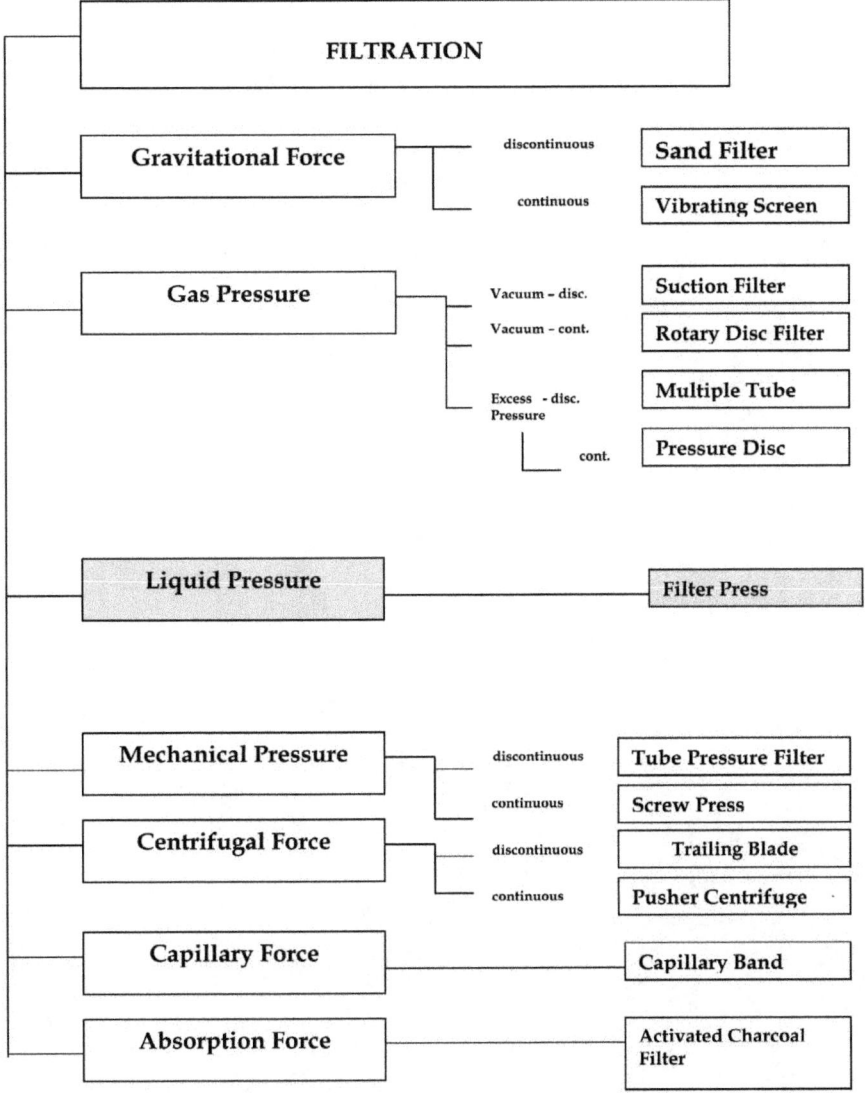

APPLICATION OF FILTER PRESSES

The question of which filter system may be used for an individual separation problem can be checked easily by the solid content in the sludge and the time to build up a cake.

Filter Devices

	throughput rate (m³/h)	solid/ content (g/l)
Basket Strainer	1000	up to 0.1
Candle Filter	300	1
Filter Press	1000	0.5 – 300
Pressure leaf filter	50	5 - 800
Vacuum drum filter	30	20 – 500
Screw Press	50	5 – 5000
Belt filter	400	100 - 1000

Cake filter Devices

	Filtration time (min)	time (min) for build-up of initial cake 10 mm
Suction filter	1-20	up to 5
Continuous filter device	up to 3	up to 10
Pressure filter device and Filter press	20 – 900	5 - 700
Sluicing pressure filter	200 – 1000+	70 – 10,000+

For an average application with filter presses the following basic characteristics of the suspension should be given:

The solid content should be in a range between 0.4 and 200 g/liter

The initial time for cake build up should be 0.02 to 3 mm cake/min.

Both values can easily be determined by a test. To find out the initial cake build up time even a simple vacuum suction filter is sufficient. However, even suspensions with values outside of the proposed ranges can be filtered successfully and economically. This is based on the extraordinary flexibility in filtration with a filter press.

The results to be achieved by filtering a suspension with a filter press can only be determined by tests with the actual suspension.

Contrary to some continuously operating filter devices, individual tests have to be done when filter presses are involved. Minimum size of a test unit should be 500x500 mm, anything below that can cause deviations of up to 30 % in comparison with later determined size of filter press to be used.

Results that are achieved with similar or seemingly similar suspensions, should be considered as a guideline only.

FILTER PRESSES
FUNCTION OF FILTER PRESSES

The filter press is a metal construction which holds the filter elements and the filter cloth.

A slurry, i.e. the product to be filtered flows through the feed inlet into the filter press and is distributed through the filter element feed port, which can be corner feed, top center, bottom center or center feed into the space between the filter elements. The space between the filter elements is formed by the corresponding cake thickness required for the liquid-solid separation. The filtrate passes through the filter cloth which is fixed on the filter elements. From there the filtrate flows through the filtrate outlets at the corner of each filter element. The solid portion of the slurry remains in the chamber between the filter elements on the filter cloth surface and gradually during the feeding of the filter press builds up a cake. This process takes place simultaneously in all chambers of the filter press. Chambers from 3 to 100 depending on the size of the filter press and required output are possible. Filter presses are designed for 7, 15 and alternatively 30 bar application.

The filter elements have to be installed in the filter press in such a way that they are tightly closed to form together with the filter cloth a sealed package through which the slurry can flow. To meet these requirements the filter press must be designed according to certain criteria.

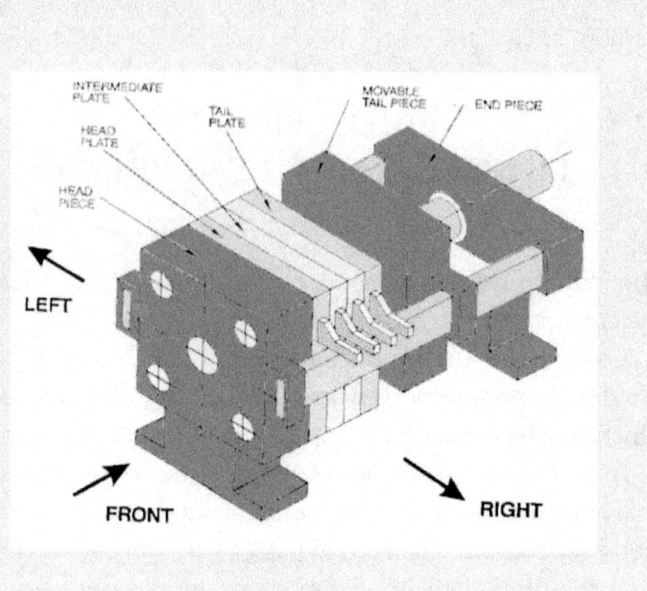

CONSTRUCTION AND DESIGN OF FILTER PRESSES

As shown in a simple sketch the filter press consists of the following major components:

1. A fixed head with the inlet and outlet connections (head plate).
2. A moving end which applies the necessary force on the filter element/filter cloth installation (end plate). The necessary force is called the closing force.
3. A cylinder end which holds and supports the hydraulic cylinder.
4. Two sidebars that balance the closing force exerted by the hydraulic ram. Other versions such as overhead suspension arrangements are possible and depend on the size of the filter press and the manufacturer
5. The installation of the filter elements and filter cloth.
6. The hydraulic power pack which supplies the hydraulic pressure to the ram.

Apart from the sidebar construction there are also overhead suspension presses in the market. The smaller filter presses, mostly those involving manually actuated cake discharge have sidebars. On the large presses the elements are freely suspended on overhead carriers; this permits full access to the plates and easier operation. The automatic mechanical plate shift system which is the norm for large presses, and which allows the full automation of the system, is located between the overhead carriers.

Designs are also available where the plates are located on side bars, as in the smaller presses, with an automatic mechanical plate shift system located below the side bar. This leaves the overhead space of the press clear where necessary for access or low headroom.

FILTER PRESS BASIC LAYOUT
(SIDE BAR DESIGN)

SIDE VIEW

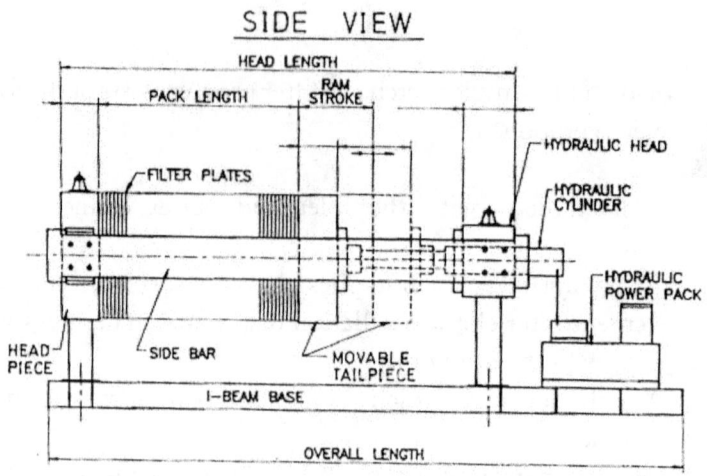

FRONT VIEW

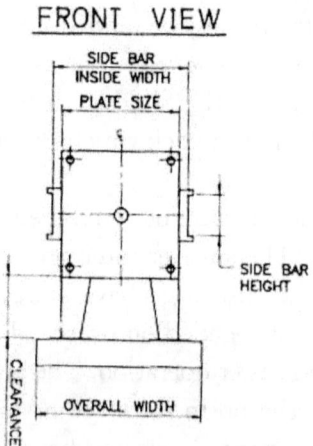

To design a filter press some calculations have to be carried out, which are similar but may vary slightly from filter press manufacturer to filter press manufacturer.

The first step is to determine the **closing force** = F (kg). This is done by the following calculation:

Overall plate area (cm^2) x filtration pressure (kg/cm^2) x 1.1 (safety factor 10%)

The above formula is sufficient in respect of filtration and squeezing pressure.

For plates and frames the closing force should be increased by 40%.

The sidebars of the filter press can be rectangular or round. The rectangular design is the most common and is used for the high loading of such a configuration, the simplicity of adding a wear strip on the top surface and the suitability of fixing the automatic plate shifter mechanism, if required. The sidebars are calculated for tensile strength of up to seven times the ultimate tensile strength of the material used.

The moving head, fixed head and the hydraulic head are of box construction design. They are designed in such a way that the closing force is evenly distributed over the entire surface of the ends. The fixed head normally consists of a thick machined face on the side that is in contact with the filter plates, reinforced with thick steel sections and boxed in with stiff plates to achieve a solid metal block.

In some cases a stainless steel lining can either be provided for the entire surface of the fixed head facing the filter plates or around the inlets/outlets to provide a corrosion resistant surface.

The moving end is similar design, the cylinder end is a box construction design suitable for mounting the hydraulic cylinder with a front flange. During operation the entire force of closing is transmitted directly from the ram to the cylinder end. The bolts on the hydraulic ram flange are therefore used only for retracting the moving head to design the strength of the head plate and the hydraulic head the total force to be transmitted through the two side bars has to be considered. In case of very long sidebars the twisting factor has to be taken into account and the sidebars have to be reinforced by two perpendicular stiffeners.

Filter presses are generally over designed for strength and rigidity reasons. Steel parts are usually sandblasted and epoxy coated. The design of the hydraulic cylinder is based on a medium hydraulic pressure of 210—250 bar. The force exerted is calculated as follows:

P (kg) = piston area (cm^2) x pressure (kg/cm^2))

For example: 280 mm diameter piston at 210 bar will produce a force of 130 tons

$$\frac{280 \times 280 \times \pi \times 210}{100 \quad 4 \quad 1000} = 130 \text{ tons}$$

The hydraulic ram will normally be of double acting design, i.e. it can move in two directions depending on the flow of the hydraulic oil through the two ports or connecting holes in the cylinder.

The hydraulic power pack will often consist of a motor, which drives a two speed hydraulic pump. The pump will deliver higher flow of hydraulic fluid at low pressure and at high pressure, i.e. during the closing mode, the flow will be reduced. This will ensure high closing speed of the filter press, which will result in a shorter cycle time. The hydraulic pack can be electrically or pneumatically powered.

An automatic plate shifter may be incorporated and is often hydraulically operated and is driven by the same power pack through a hydraulic motor, which turns the sprocket wheels. The two sprocket wheels are connected with a common shaft, which will ensure synchronizing movement of the plate shifter. The reversing of the plate shifter is initiated by a pressure sensor. The advantage of a hydraulically operated plate shifter is the variable travel speed and the variable force for moving the filter plates.

The above basically describes the design and the construction of a filter press. As previously indicated there are different designs and constructions in the market, which depend on the filter press manufacturers' specific criteria.

Filter presses can be further equipped with PLC control; Bombay doors, safety features and flow control as well as filter cloth cleaning systems.

The PLC controls all the valves automatically for the feed cycle and the cake washing cycle and opening and closing of the press. The filter press can then operate with minimum attendance of the filter press operator. Valves are pneumatically operated from the PLC control.

Bombay doors are installed to collect any remaining liquor that is left in the chambers between the filter elements while the closing pressure is released. Bombay doors are made from a variety of materials,

LIQUID SOLID SEPARATION WITH FILTER PRESSES

which can be stainless steel or aluminum. They are normally operated hydraulically.

Safety features are important to prevent the filter press is being opened during a feeding or squeezing process. An interlock system is installed to prevent the release of hydraulic pressure while squeeze pressure is applied to the membrane.

PROCESSING IN A FILTER PRESS

A filter press cycle, especially with a membrane filter press, is divided into several steps until the filter cake is discharged and a new cycle starts. The following steps run after starting the cycle:

- closing of the filter press
- opening and/or closing of the valves
- pre coating of the filter cloth
- running the feed pump for filtration of the suspension
- pre-squeezing of the membrane
- washing of the filter cake
- final squeezing of the filter cake
- cake drying by air
- core blowing
- discharge of the filter press chambers
- closing of the filter press

OPTIMIZATION OF A FILTER PRESS CYCLE

When a filter press cycle has to be optimized, all the above steps have to be checked. Generally, all the required steps in a cycle can be optimized in two ways. In practice it is possible to look for more economic efficiency and/or higher throughput of the filter. Very often both improvements are connected with each other. The efficiency of the processing of the product is very individual and will not be discussed here. However, the optimization of the filter press cycle in order to increase throughput and reduce labor costs is much more general with regard to all kind of applications. Essentially, there are two kind of process steps; filtration which is always the only productive time, and the remaining time which is non-productive and added together as downtime. The filtration time of a chamber filter press is mostly a question of the pump size, the kind of slurry treatment and the chamber depth, and if everything runs perfectly then not much time and cost saving is possible. The use of a membrane filter press offers more possibilities. The final dry solids content of the filter cake is reached by inflating the membranes and squeezing the moisture out of the cake. Therefore filtration is possible at the most economic point in time and with the best input.

MINIMIZING OF DOWN TIME

All steps in the filter press cycle, with the exception of the filtration time, are added together to give the downtime. Usually the discharge time is a rather large component of it. Depending on the press size and the filterability of the product, the total downtime could be as little as only 10% or more than 50% of the cycle time.

DISCHARGE TIME

The time taken to discharge a filter press also varies within a wide range. Small filter presses where the filter cake falls off very easily are discharged in a few minutes, even when everything is done manually. However, larger filters with a greater number of chambers take much more time to discharge.

Discharge of a 1500 mm square press with 100 chambers, with the filter cake slightly sticking to the cloth so that the operator has to control the cake discharge manually, takes about 1 hour. This is only 30 seconds per chamber. In many applications this time can even account for more than 50% of the cycle time. It is very simple to calculate how the throughput increases when the discharge time is cut down to a few minutes. In addition, there is a lot of labor saving when discharging runs automatically. These facts and circumstances suggest that filter press operation would be improved by an automatic cake discharge system.

MOBILE FILTER PRESS

Mobile filter presses are filter presses of the same design, mounted on a trailer or a truck. The only limitation imposed is that of the traffic regulations of the particular country. Virtually every country limits the weight, length and width of vehicles using public roads. Apart from the filter press the installation of mixing and holding tanks and the required compressor to supply the squeezing media for the membrane filter elements has to be taken into consideration.

Mobile filter presses are mainly used for on-site waste handling. These units are mainly used where chemical process companies have collected a large volume of waste, often hazardous, in lagoons, because the daily amounts of waste are small it is not economic to install a permanent waste treatment plant and lagooning makes it possible to collect sufficient quantity to make it viable to bring to site the mobile waste treatment equipment. The availability of filter presses is of special interest to those companies that have to treat the waste once or twice a week normally due to the small amount of sludge on a small filter press. Although a small filter press does not involve high investment costs it can be advisable in some cases to entrust companies that are specialized in the on-site waste treatment.

These specialized companies do not only provide a filter press and the necessary manpower to filter the sludge; they also provide the materials that are needed on site. These specialized companies charge based upon slurry volume, dry solid weight, daily etc. Mobile filter presses also contribute to a reduction of the quantity of solid wastes, which normally would have to be transported to permanent off-site facilities for disposal.

Some data on a possible mobile filter press design:

Weight (kg)	8000 – 40.000
Total width (mm)	1800 – 2.500
Total height (mm)	1.600 – 3.500
Total length (mm)	8.000 – 14.000
Filter element size (mm)	470 – 1500
Number of chambers	up to 90
Pump capacity m3	10 – 50

The above figures are for the trailers only.

Depending on the size of filter press and overall capacity a mobile filter press can easily cost up to US $ 500,000.

This amount includes trailer, filter press with semi-automatic plate shifter, belt conveyors (extendable), cake disposal system (chutes), manifold for cake washing, air blowing, railing system, stairs, walkways, compressors, tanks, pumps, illumination system, control panel, squeeze system for membrane filter elements.

FILTER PRESS ACCESSORIES

Closing device

Manual closure

Manual closure is mainly used for filter presses up to size 630. A hand operated hydraulic jack is applied to manually close the filter press.

Automatic closure

Automatic closure is done by air-operated hydraulic systems. The size of such a device corresponds to the surface area for the filtration and the pressure of the feed pump. The plates are automatically pressed together.

Blanking Plate

A blanking plate or back-up plate can be inserted at any point in a plate stack in order to isolate all the filter plates between the blanking plate and the follower.

With this method less filter plates are used for filtration than actually provided in the filter press. The end plate has to be relocated and inserted in front of the blanking plate in order to provide a complete chamber with the remaining filter plates. This method is used in case where the amount of product to be filtered varies.

FILTER CLOTH WASHING

Since the filter cloth is to separate the solids from the liquid is has to be cleaned in order to remain porous and in order to provide a high filtration rate.

During operation the filter cloth becomes plugged. Particles enter the cloth and are trapped in the depths of the weave, which causes a decreased filtering process.

To ensure proper operation these particles must be removed.
Typical signs for such situations are:
Wet filter cakes
Higher filtration pressure
Long filtration cycles
There are several methods available to wash filter cloth.

1. Small compressor to operate at 70—110 bar
2. Acid washing through the press
3. Dip-washing of the plates

1. Simple operation. Not only the surface of the filter cloth is cleaned but also penetrated to flush the particles out of the depth of the weave.
2. An acid storage tank is required with a capacity of 1.5 times the holding capacity of the press. Depending on the solubility level of the particles a 25% solution of hydrochloric acid is sufficient. The execution of the pump should be acid-resistant to run at 1-2 bar. Proper piping to isolate the filter press from the sludge stream allowing re-circulation to the acid storage tank and final draining has to be arranged for by the filter press manufacturer. Before the cleaning is initiated the remaining cake on the filter cloth has to be removed.
3. Dip washing is a time consuming procedure. The plates are immersed in an acid tank. This method does not provide the efficiency as through washing. Overnight soaking to clean out

the depth of the weave is mostly necessary. The plates due to the weight will float since they are lighter than water. Steps are necessary to keep the plates submerged.

FILTER CLOTH WASHING DEVICES

1. Gap washing device
2. Plate washing device

Gap washing device

The moment there is enough space between two plates and the right positioning of the plates is reached a tube with spray nozzles is moved into the space between the two plates and starts cleaning the cloth.

Plate washing device

With a plate washing device the filter plate is cleaned on both sides at the same time. In comparison to the gap washing device the plate washing device can work at 70 bar achieving the same results as the gap washing device at 100 bar. It has to be considered in case of the plate washing device that the filter press has to be equipped with a connection plate since the head plate will be cleaned too on the backside. The plate washing device offers the advantage that the total height of the construction is lower than for the gap washing device.

HANS G. VALERIUS PHD

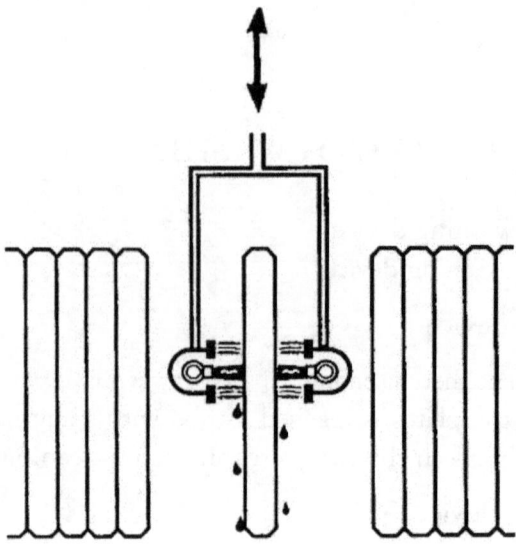

AUTOMATIC CAKE DISCHARGE

The filter cloth is hooked on top and by moving the plates the position as per illustration is achieved, discharging the cake easily. For this kind of automatic cake discharge the positioning of the corner infeed or center infeed must be in the lower part of the filter plate, i.e. center feed—bottom center.

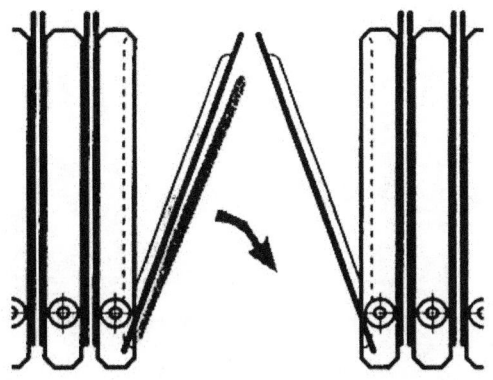

CAKE REMOVAL DEVICES

Dumpster

For handling of discharged cake dumpsters can be provided. These dumpsters are mounted on ball bearing casters for easy positioning under the filter press. When the cake is dropped into the dumpsters same can be removed manually or by forklift truck, positioned over a larger container and emptied by pulling a spring-loaded self-dump bin.

Drum disposal system

The drum disposal system is used in connection with filter presses, which are mounted on an elevated platform. Chutes between the filter press and the drums make sure that the cake is properly dropped. A so-called drum cart is provided to remove the drums after being filled with the cake.

Chutes

In cases were dumpsters or drums are not used the press can be installed on extended legs to provide a barrel type disposal. Chutes have to be provided to make sure that the cake is directed into the collection barrel.

Roll-off container

Also here the filter press is mounted on a platform to provide sufficient space for the container. A large-sized chute under the filter press directs the cake into container.

Drip tray

A drip tray contains overlapping sheets fixed on a metal frame. The reason for having such a design is to collect liquid coming out from between the plates. The tray is slanting off to one side where the liquid is collected in a launder. Before the filter press is opened the trays have to be removed in order to provide free space for the cake discharge.

Bombay Doors

Bombay doors an equivalent to drip trays are Bombay doors. Bombay doors are in automatic system of the drip trays. Before the filter press is opened for cake discharge the doors are automatically opened and put into vertical position.

Conveyors

Another method of cake discharge is the use of a conveyor. When conveyors are applied the distance between the filter press legs has to be considered. In some cases spread legs might be required.

SAFETY DEVICES ON FILTER PRESSES

Safety guard

On the non-operating side of a filter press a metal screen is installed to avoid that the press is approached during opening or closing.

Light curtain

The light curtain is a safety feature placed on the operating side of the filter press. Infrared beams between the sender and the receiver sense any kind of penetration of the beam by any object. The moment any object is crossing the beam, the operation of the press is interrupted until the object is removed. The light curtain will not operate once the plate stack is in the intended position and the required hydraulic pressure is achieved.

Splash curtain

A splash curtain can be installed around the filter press and help to contain liquid which could splash from between the filter plates during filling or washing procedures.

SIZE AND CAPACITY OF FILTER PRESSES

The following calculations are guidance on how to determine the size of a filter press:

$$\frac{\text{Total sludge volume per day} \times \text{spec. gravity of sludge} \times \text{solid content of sludge in \%}}{100} = \text{total dry solids per day}$$

$$\frac{\text{Total dry solids per day} \times 100}{\text{Solid content in cake in \%}} = \text{total weight of filter cake per day}$$

$$\frac{\text{Total weight of cake per day} \times 100}{\text{Spec. gravity cake}} = \text{total volume of filter cake per day}$$

$$\frac{\text{Total volume filter cake}}{\text{No. of cycles}} = \text{total volume per cycle}$$

$$\frac{\text{Total volume per cycle}}{\text{Volume of chamber}} = \text{total number of chambers}$$

$$\frac{\text{Total no. of chambers}}{\text{Max. no. of chambers per filter press}} = \text{total number of filter presses}$$

The volume of the corresponding filter element can be taken from the data sheet which is provided by the filter element manufacturer. See sample on next page (Chamber plate with various cake thickness, indication of filtration area and volume).

As a reminder test filtration is necessary. The above is only a guidance or can be applied when experience with the corresponding product is already available.

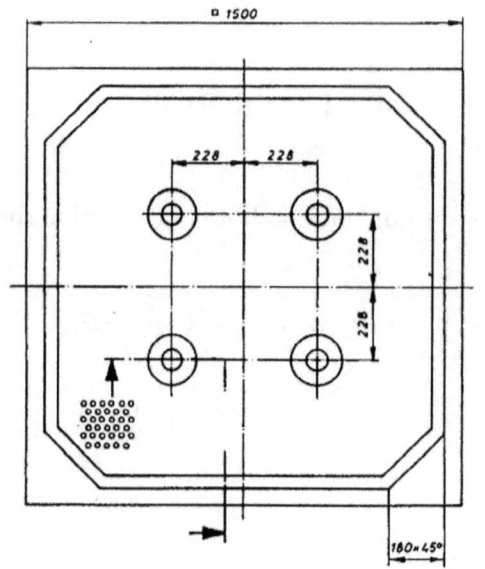
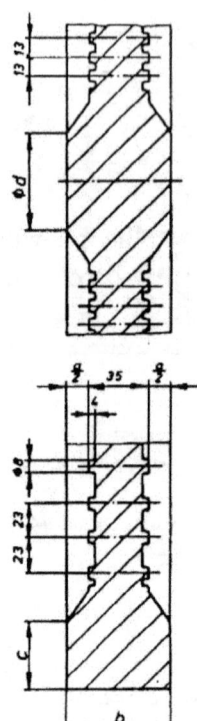

cake-thickn. a	plate thickness b		sealing-edge c	distance-support ⌀d	filterarea m2	volume liter	weight kg
	chamber plate	headplate endplate					
35	70	62.5	64	96	3.574	59.7	83.6
38	73	64	62	92	3.601	65.1	84.8
40	75	65	61	89	3.619	68.7	85.6
45	80	67.5	58	82	3.664	77.8	87.6
50	85	70	55	75	3.708	86.9	89.5

HOW TO SIZE A PUMP

There are various methods of calculating the size of a pump:
1. Filter Press Volume x 12 = Feeding Capacity/hr
2. Filtration Area x 225 ltr/m2/hr

Example 1: Volume 7065 ltr x 12 = 84780 ltr = 85 m3/hr

Example 2: Filtration are 477.1 m2 x 225 ltr/m2/hr = 10.7347 ltr = 107 m3/hr

The above serves as example only. For palm oil a figure of 350-400 ltr and for kaolin a figure of 300 ltr should be taken.

APPLICATION OF FILTER PRESS—GENERAL

Filter presses can be seen in operation in may industries, like:

Food and Beverage Industry

Whiskey	Fruit Juices
Spices	Margarine
Protein (H.V.P.)	Soya Sauce
Pactin	Seaweed
Palm Oil	Soups (dry)
Wine yeast	Beer yeast
Edible Oil & Fats	Rice Starch
Bovril	Coconut Cream
Beet Sugar	Cane Sugar
Bakers Yeast	Palm Kernel Oil

Chemical Industry

Dyestuff	Dyestuff (Pigment)
Intermediates	Basic Chemicals
Filling material	Titanium Dioxide
Silica- Calcium	Calcium Carbonate
Carbonate-Zeolite	Glycerin
Catalysts	Calcium Hydrochloride
Arsenic Sulfite	Amino Acid
Enzymes	Sodium Chloride
Antibiotics	Soap
Sodium Hydrosulfite	Sulfuric Acid

Metallurgical Industry

Nickel-Sulfate	Sodium Sulfate
Alum-Hydroxide	Barium-Hydroxide
Aluminum Oxide	Manganese-Sulfate
Zinc-Chloride	Lead Phosphate
Manganese Chloride	Iron-Hydroxide
Aluminum Hydroxide	Stearates
Potassium	Potassium Chloride

Minerals-Earth Industry	Bleaching earth	Bentonite
	Tixogel	Kaolin
	Ceramic Slip	Cement
	Fine earth	Industrial sands
Mining Industry	Gold	Platinum
	Uranium	Cobalt
	Nickel	Zinc
	Coal fines	Alumina/Aluminum
	Ironware	Phosphate
	Manganese	Copper
Waste Water	Industrial	Municipal
	Galvano	Coal Waste

GENERAL SLURRY TESTS WITH FILTER PRESS INCORPORATING MEMBRANES AS GUIDELINE 1 :

Slurry	Moisture %	Other systems
Chemical industry		
Aluminum silicate	65	82
Barium sulfate	19	21
Lead phosphate	26	-
Lead sulfate	18	-
Lead sulfite	24-30	-
Calcium carbonate	21-42	60
Calcium sulfate	12-25	-
Chromium oxide	16	25
Iron oxide	14-54	-
Hydromagnesite	58-60	62
Silicic acid	65-75	80
Optical brighteners	40	-
Stearates		
Aluminum	43-50	52-59
Calcium	32-44	52
Lead	22-25	-
Magnesium	37	50
Zinc	38-44	44-55
Titanium oxide		
Before calcination	50-53*	-
After calcination	34	50
Tripolyphospate	65	75
Zinc carbonate	43	-

*drying at 900°C

Actual results depend upon the slurry feed condition, the particle size and the process itself.

Actual results depend upon the slurry feed condition, the particle size and the process itself.

Slurry	Moisture % Membranes	Moisture % Conventional	Savings Dryness	Washing	Handling
Food & Pharmaceutical					
Fermentation broth	15	35	x	x	x
Wheat starch	37	45	x		
Yeast	50	60	x		
Protein	36	55	x	x	x
Corn gluten	48	60	x		x
Waste					
Scrubber slurry	15.5	30	x		
Fiber sludge	47	70	x		x
Metal Hydroxide	50	80	x		x
Wood fiber	49	65	x		x
Dust	16	31	x		x

LIQUID SOLID SEPARATION WITH FILTER PRESSES

Slurry	Moisture % Membranes	Moisture % Conventional	Savings Dryness	Washing	Handling
Concentrate Processing					
Gold concentrate	7	14	x		x
Copper concentrate	4.5	9	x		x
Molybdenum concentrate	7	15	x	x	x
Nickel concentrate	7.5	15	x		
Lead concentrate	5.1	10	x		
Zinc concentrate	5.5	11	x		
Industrial Minerals					
Talc	7.9	16	x		x
Calcium carbonate	10	20	x		
Precipitated CaCO3	16	35	x		x
Apatite	5	11	x		x
Coal	12	21	x		x
Metallurgical					
Aluminum hydroxide	30	45	x		
Aluminum oxide	27	40	x	x	
Aluminum sulfate	30	50	x		
Slag	8	17	x		x
Anode slime	10	22	x	x	x
Zinc sulfate	14	30	x		
Chemicals Processing					
Silicate pigment	19	30	x	x	
Magnesium hydroxide	17	32	x	x	
Catalyst	74	90	x	x	x
Sulfur	10	24	x		x
Polymer	19	35	x	x	x
PVDC	24	40	x		

Slurry	Moisture %	Other systems
Food Industry		
Corn glue (starch)	45	55
Rice flour	44	49
Soup flavors	27	-
Citrus chips	57	-
Sugar production:		
Carbonation juice	25-30	-
Metallurgical Industry & Mining		
Shaley water	20-22	22-26
Copper concentrate	8	-
Lead concentrate	5-7	-
Zinc concentrate	5-14	8-16
Magnesium hydroxide	29-33	37-40
Quartzite ore (gold)	12	-
Crude tungsten acid	19	-
Pollution Control Water Treatment		
Calcium carbonate	30-40	n/a
Calcium chloride	37	n/a
Calcium sulfide	42	n/a
Ore-coal dust	21	n/a
Carbide sludge	37	n/a

NECESSARY DATA FOR SIZING THE FILTER PRESS

1. Product
2. pH-value
3. Operation temperature
4. Operation pressure
5. Infeed pressure
6. Squeezing pressure in case membrane filter elements are involved
7. Turbid volume to be filtered daily/per cycle
8. Dry solids to be separated daily/per cycle
9. Total filter cake volume
10. Operation time total
11. Cycle time total
12. Cycle time split into the individual steps like feeding, filtration, squeezing, washing, core blowing, cake discharge, etc.
13. Number of cycles per day
14. Filtration task, interest in solids or filtrate to be retained
15. If washing is involved, media for washing, quantity, pressure, temperature, pH-value to be achieved
16. Desired dry solid content
17. Specific Weight
18. Structure of product, whether glutinous, colloidal, granular, crystallized etc.
19. Whether product is toxic
20. Viscosity

Not all the above details can be given, however the scope of information should be as comprehensive as possible.

NECESSARY TEST FILTRATION

For test filtration purpose small filter press units usually size 500 x 500 or 630 x 630, are used. The results achieved in the course of a few tests are recorded to give the basis for the decision on the most economic process. The result from the test filtration with the aforementioned units can be scaled up to the full sized filter press. Various filter cloth types and chamber depths have to be tested.

Test filtration results with small test units give good data. It will always be necessary to carry out final "tuning" of the full size press as it is not possible to achieve an exact scale-up of every component of a filtration system.

Results of a test filtration could be as follows:

	Time (min)	Pressure (bar)
Filling	15	3
Pre-squeezing	2	3.5
Washing	30	4
Squeezing	5	5
Core washing	1	1
Core blowing	0.5	2
Chamber blowing	0.5	1
Discharge	30	
Totals	84 mins	

Time (min)	84
Slurry weight (kg)	115.42
Cake weight (kg)	30
Cake dry solids content (% wt/wt)	25
Cake thickness (mm)	30

Test Press

Cake thickness	Chamber volume dm3	Set up
20mm	10.84	1-3-1
25mm	13.55	1-2-1
30mm	16.26	1-4-3-4-1
35mm	18.97	1-4-2-4-1
40mm	21.68	1-1
50mm	27.12	1-4-1

Given : Customer needs a slurry throughput of 10.000 kg per hour
Test press filtration area (m2) = 1.124

$$\text{Throughput} = \frac{\text{slurry weight (kg)} \times 60}{\text{Total time (min)} \quad \text{Total area (m2)}} = 73.348 \text{ kg/hr/m2}$$

$$= \frac{115.42 \times 60}{84 \quad 1.124} = 73.348$$

Filtration area required = $\frac{\text{Customer Requirement (kg/hr)}}{\text{Throughput (kg/hr/m2)}}$

$\frac{10.000}{73.348} = 136.34$ m2

Arrangement of filter elements for test filtration should be as follows in order to allow flexibility during test filtration.
1. 3 Membrane (2 membranes serve as head and endplate)
2. 2 Chamber plates
3. 2 Filter plates
4. 4 Frames

Arrangement Variations for testing (the set up configuration refers to the numbers above, i.e. 1 = membrane, 2 = chamber plate and so on.

LIQUID SOLID SEPARATION WITH FILTER PRESSES

Cake thickness	Chamber volume dm3	Set up
20mm	10.84	1-3-1
25mm	13.55	1-2-1
30mm	16.26	1-4-3-4-1
35mm	18.97	1-4-2-4-1
40mm	21.68	1-1
50mm	27.12	1-4-1

TROUBLE SHOOTING

In many cases filter press operators will state that the filter plates are faulty with little evidence to support the statement. The following subjects refer to general statements and the possible reasons:

Leakage from the plate edges

- Insufficient closing load
- Fold in filter cloth
- New filter cloth
- Build up of cake on sealing edges
- Plates bend so that sealing faces do not lie flat on each other
- Damaged sealing faces
- Press over-packed (this can occur even when correct closing force is applied)
- Cloth misaligned at corner eyes
- Seals of connection/Adapter plates damaged
- Seals of caulked and gasketed plates are damaged
- Insufficient Slurry
- Incorrect cloth allows solids to pass
- Damaged cloth allows solids to pass
- Misalignment of cloth between corner eyes and chamber allows solids to pass

Wet cake

- Insufficient pump pressure
- Blinded Cloth

Bent plates/thick/thin cakes

- Improper filling
- Plugged drain ports
- Plugged feed ports
- Blinded cloth

- Drainage surface blocked
- Mixture of different cloth types

Poor Washing

- Poorly formed cakes
- Insufficient wash water
- High low wash rates
- Plates installed incorrectly

For Membranes

Hinge area white (normal because the molecules re-align under stress) Bulging membranes (often occurs with age or over stretching)

Will not hold pressure : Check all connections first
Grommets missing
Cloth over grommets
Split membrane

Tracing a Leaking Membrane

Grommeted squeezing systems make tracing a leaking membrane a lengthy process. Each plate has to be tested individually up to not more than 0.25kg/cm2 by applying air through one grommet hole. Soapy water over the plate or complete submersion can help.

With the external squeezing system it is easy to remove the hoses from the plates. Then the press can be filled under pressure and liquor will come from the connection of the plate that is normally used for the squeezing media.

LIQUID SOLID SEPARATION WITH FILTER PRESSES

Some recommendation

Plate leakage	**To be done**
Plugged feed ports	Remove same
Plugged drain ports filter cloth	Remove same, also check on (some filter cloth allows more solids to pass though. Check the plate drainage behind the cloth)
Pump is irregular	Check pump and press. Start again at low pressure and slowly build up pressure.
Short cycle time and sufficient solids.	There is not enough slurry. Insert back up plate.

High velocity pumps used as feeding pump might cause leakage due to velocity pressure on only one side of the plate.	A pump with pressure/flow curve identical to the filtration curve should be used.
Wrong application of plates	Avoid mishandling. Avoid that plates are dropped on side bars
Wrong application of back-up Plates.	Position back-up plate behind endplate.
Unclear filtrate	**To be done**
Damaged filter cloth	Check same
Caulking out of grooves	Replace caulking (caulked & gasketed version)
Filter cloth is pulled out of groves **(Caulked & gasketed version)**	
A full cake was not built up before washing	To avoid that the cloth is pulled out of the groves a proper cake built-up is necessary otherwise the cake cannot support the cloth.
The cord (O-ring)/or the filter cloth is not properly sized	Change size of future cloth

MAINTENANCE IN GENERAL

Daily

- Clean sealing areas of solid build up.
- For caulked & gasketed version clean area and replace with signs of damage, i.e. cuts etc.
- Check on filter cloth damages, like holes etc.
- Check on any leakages.

Weekly

- Check on oil level in hydraulic reservoir
- Check on relief valve setting
- Washing filter cloth depending on process (with caustic or acid)

Monthly

- Clean oil filter
- Wash filter cloth depending on process (with caustic acid)

Yearly

- Replace oil filter
- Replace oil in hydraulic system
- Replace any gaskets if any

FILTER ELEMENTS FOR DISCONTINUOUS SEPARATION SYSTEMS

Filter elements are applied for liquid solid separation of cake building systems.

With all cake building separation systems, either continuous or non-continuous, the filtration shows individual process steps that are explained in the following chart.

STEP 1 FEED OF SUSPENSION INTO THE CHAMBERS
STEP 2 CAKE FORMATION
STEP 3 DEWATERING OF CAKE (IF REQUIRED CAKE WASHING)
STEP 4 CAKE DISCHARGE

The difference between continuous and non-continuous systems is given by the sequence of the steps referring to time and to the position inside the filter unit.

With a discontinuous system, contrary to a continuous one, unavoidable dead times, for example for cake discharge or cleaning of the system, occur. On the other side continuous systems do also have unavoidable interruptions, i.e. for maintenance.

Continuous systems require not only more maintenance but also availability of wear and tear parts which is reflected in additional costs, i.e. belt presses. Nowadays discontinuous systems can be more or less automated that the actual difference between the advantages or disadvantages of a continuous system vs. a discontinuous one can be neglected.

The discontinuous process however, allows adjustment of any single process step, i.e. cake build-up or subsequent wash cycles, and allows later changes in a wide range.

The discontinuous filtration system can be adjusted for an individual separation problem very easily. Suspensions or sludge which are extraordinarily difficult to separate and that vary in their characteristics can be processed due to the flexibility of the filtration system. A continuous filtration system however, cannot be easily readjusted to handle changed characteristics of a suspension without expensive changes in design because the sequence of steps in time are fixed by the construction of the filtration/separation system.

FILTER ELEMENTS FOR FILTER PRESSES

Today three categories of filter elements are applied in filter presses. These are:
1. plates and frames
2. recessed chamber plates
3. membrane plates

All these types of elements build a room (chamber) for the separation of the liquid from the solids.

The filter elements are kept together in a hydraulically closed device, which is the filter press. Chambers are formed between the filter elements. The cross section through a typical filter press equipped with recessed chamber plates shows the basic arrangement of filter elements and the assembly of the system.

The main difference between the three categories of elements is found in the way they form the separation chamber.

FILTER ELEMENTS WITH FIXED CHAMBER DEPTH

The oldest design of filter elements are filter plates and frames. The filter plates are flat and form the separation room (chamber) when a frame is inserted between two of them. The depth of the filter chamber is fixed by the thickness of the frame and varies from product application to product application

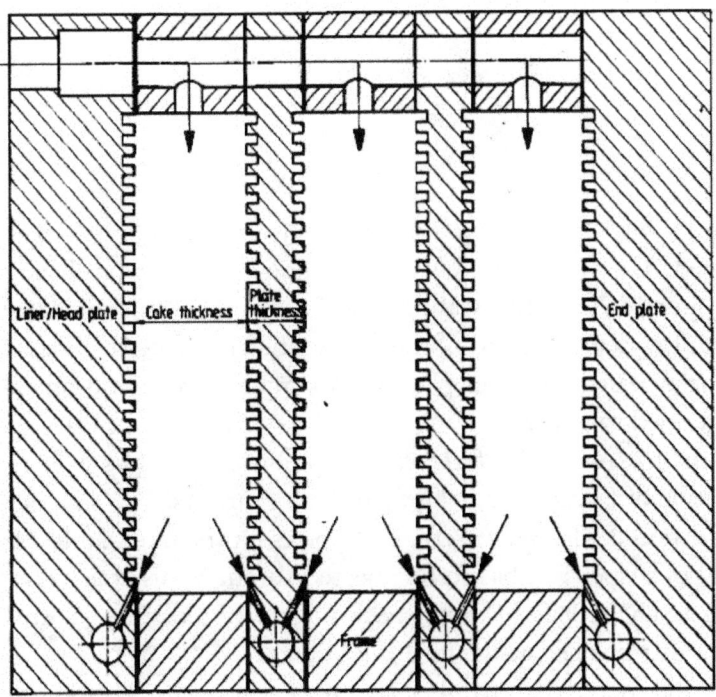

An advantage of the filter plate and frame arrangement is that they exert only low forces on the filter cloth, consequently inexpensive overhanging filter cloth as well as filter paper can be used.

Moreover, the chamber depth can be changed easily by replacing the

frames with frames of another thickness, an interesting aspect for these companies that have a wide product range.

Due to the arrangement of plates and frames and the construction itself, automatic cake discharge is not possible.

From filter plates and frames the recessed chamber plates were developed. They are, unlike filter plates, no longer flat but have a raised outer edge which is called sealing area. The height of this edge determines the chamber depth instead of the frame, thus the cake thickness.

The recessed chamber plate is in simple words—a filter plate and frame welded together.

Over this edge the filter cake slips off easily, thus a first step to automation of filtration is done with these elements.

The raised edge now exerts high forces on the filter cloth. Moreover, the feed of suspension has to be injected through this edge. This means that relatively expensive barrel neck filter cloths or filter cloth locking devices must be used at the feed inlet ports.

Furthermore, the chamber depth is fixed during production and cannot be changed later except by replacement of the recessed chamber plate.

The recessed chamber plates are, due to their simple design, the least expensive filter elements since the frame is no longer required.

SIMPLE FILTRATION OF CHAMBER PLATE OPERATION

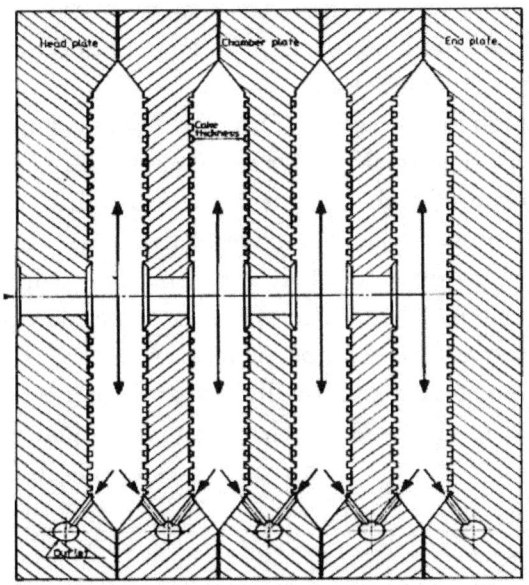

Caulked and Gasketed Filter Elements

Apart from the standard recessed chamber plate with different forms of filtration area, there are also available so-called caulked and gasketed recessed chamber plates.

These plates are used for leak-free operation and especially in production sites where the product handled has a strong smell or is hazardous in nature.

The construction of the caulked and gasketed recessed chamber plate does not differ very much from the standard version.

The difference is in the seal around the edge of this filter element and the groove around the filtration area where the filter cloth is secured.

Filter elements with variable chamber depth

Filter plates and frames as well as recessed chamber plates have, due to their design, a fixed chamber depth.

Membrane filter plates look similar to recessed chamber plates, however the body between two chambers is constructed of several layers. The layer adjacent to the filter chamber, the membrane, is designed as a diaphragm however, has nothing in common with same. When a pressure is built up behind the membrane, it balloons inside the filter chamber and decreases the chamber depth.

The very high forces exerted on the filter cloths by membrane plates, requires that only high quality filter cloths can be used. Additionally, the design of the membrane plates leads to a reduced chamber volume when compared to a recessed chamber plate of the same size. In addition, a squeeze system is required to inflate the membrane, which depending on the media, air or water or product will cause some additional investment costs however, the variable chamber depth provides exceptionally high dry solid contents in the cake and usually leads to a higher throughput rate. The comparison of the different element types explains the advantage of the membrane plate which in many cases results in a fast return on investment.

COMPARISON

Flush Plates & Frames	Recessed Chamber Plates	Membrane Plates
Low forces exerted on filter cloths even paper can be used	High forces exerted on filter cloth. High quality cloths are required. Paper difficult to use	Very high forces exerted on filter cloths. Thus high quality is required. Paper cannot be used.
Chamber depth can be changed by replacing frames with different thickness	Chamber depth is fixed and cannot be changed afterwards. Good cake discharge	Flexibility chamber depth due to principle of membrane plates.
Difficult cake discharge.	Less expensive to manufacture.	Excellent cake discharge.
Cake often remains inside the lower frame area.	Filtration pressure up to 16 bar	Highest investment cost due to the necessity of a squeeze system. Does not require high filtration pressure due to flexible chamber depth. Dry solid content is adjusted by membrane squeezing. Reduced chamber volume. Short cycle time, high filtration rate and low moisture content.
More expensive than recessed chamber plates due to the frames being required.	Due to occasionally blocked channels core blowing required	
Filtration pressure must be reduced in increased chamber depth.		

Membrane plates

In general all membrane plates consist of a fixed centerpiece, the web and the flexible membranes. Because of different requirements of the web and the membranes the material has to be chosen accordingly. The one and only function of the web is to give the filter element stability.

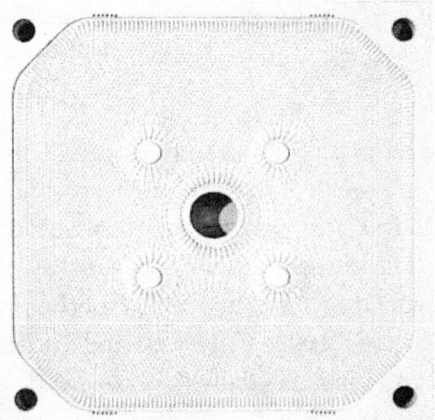

Material

Steel, ductile iron or aluminum webs.

The materials have a comparatively high tensile strength and can consequently take comparatively high differential pressures with little bending. If, however, the differential pressures exceed the force the web can take, the results will be a permanent bending with steel or aluminum plates or breakage of the ductile iron because of the material's low elasticity. Because of the weight of these metallic materials, the costly corrosion protection measures the risks of damaging the anti-corrosion lining in the daily application, filter press users are switching over to the use of the thermoplastic PP material for the webs of the membrane filter elements.

Polypropylene (PP) webs

Advantages are low specific weight (0.92 g/cm2) and thus easy handling. Excellent chemical resistance and thus no corrosion problems. Consequently no corrosion protection like lining necessary. PP being a thermoplastic material possesses high mechanical tensile strength values and a high elasticity combined with a very good "memory".

This means in practice that if the web is bent because of differential pressure no permanent deformation of the web is caused once the pressure has been released.

Design of the membrane plate

There are basically three designs: welded one-piece version
   detachable 3-piece version

The welded version is of one piece only. The web and the two membrane halves are welded together to form one unit. This version is the most trouble-free and easy to handle while the investment costs are relatively low. The three-piece version consists of the web and two membrane halves which are attached to each other by a plastic clip and a sealing ring, both are between the web and the membrane halves or it consists of 2 membrane halves which are fixed at the sealing area, the overhanging membrane. This version is recommended for use in industries where the membranes have to be replaced quite frequently due to high amount of cycles. Since the three-piece version is more expensive than the welded version the decision on whether to select the one-piece or three-piece has to be carefully considered. Some version of the detachable membrane do require maintenance like re-tightening bolts at the center feed ring and avoidance of product to enter into the small gaps between web and membrane halves.

Three-piece membrane spares, i.e. halves should be kept in stock properly supported and at ambient temperature. Wrong handling and storage can lead to problems at the time of replacing membrane halves.

LIQUID SOLID SEPARATION WITH FILTER PRESSES

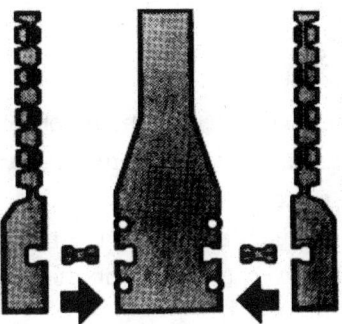

Membranes are also available in caulked and gasketed version, however, for construction reasons, mainly in the one-piece welded version. Some suppliers do have detachable membranes in caulked and gasketed version.

Because of the design of the three-piece plate, variations in material of the web and of the membrane halves are possible.

Construction of some detachable membrane plates can be rather production-intensive and consequently result in higher investment costs for the complete membrane as well as for the membrane halves than for a one-piece membrane plate.

The so-called overhanging membrane is a very simple detachable membrane, easy to operate and to replace its application however is limited due to the simple design.

Mixed pack

In some applications the arrangement of alternate membrane and chamber plate can be used. This arrangement is less costly in comparison with a full membrane plate set.

Similar results can be achieved with the mixed pack as with the full membrane plate set, however, a few points have to be considered before deciding to use a mixed pack:
1. The squeeze of the membrane applies force to the cake on one side only since the opposite side, the chamber plate, is not inflatable.
2. If chambers are not properly filled during the filling of the filter press, membranes can easily be overstretched during squeezing, i.e. by applying air/water for the squeezing the membrane is inflated and moves into direction of the cake. If no cake is built up, the membrane continues to move until the movement is stopped. Since no cake is available the movement of the membrane is only stopped by the chamber plate. The membrane is not equipped to make such movements continuously and can be damaged by overstretching.
3. There are of course systems available that reduce the risk of overstretching, however, these versions, when used for a mixed pack, do not always allow a conversion to a full membrane plate set should results show that a decision for a full membrane plate set would have been the better one. The available system is known as the empty chamber design. This design due to a slightly under sealing area of the chamber plate blocks the membrane movement towards the chamber plate surface. The initial point of overstretching is nealy eliminated. It is important to consult the manufacturer due too the variety of possibilities to avoid overstretching of the membranes, i.e. installation of distributor rings etc..

Differential pressures

Differential pressures play an important role in the conventional filtration, i.e. when using plates and frames or recessed chamber plates. In order to achieve high dry solid contents in the cake high filtration pressures are required. As a rule differential pressures between the chambers only occur when the cake is being packed by the pump pressure, i.e. in the last phase of filtration.

It is very unlikely that this will happen during membrane filtration. Contrary to the conventional filtration system the slurry pump feed pressure is drastically reduced. The dry solids in the filter cake are achieved by squeezing the cake. During the squeeze the plate pack pressures are in balance since the forces are equal in each and every chamber. However, it is important not to overfill the chambers during filtration.

Material for filter elements

Decades ago, filter plates and frames were manufactured from expensive hard wood. Due to a constant increase in costs for wood, environmental protection and the cracking of wood when allowed to dry, this material was replaced and recessed filter plates made of cast iron were introduced.

The poor chemical resistance of cast iron requires expensive lining of the filter elements with rubber or, now more common, thermo-plastic material. Furthermore, this lining has to be redone periodically. The high weight of the elements requires very sturdy filter presses and building foundations, not to forget the manpower to handle these heavy elements.

Therefore, in the last twenty years, filter elements for filter presses have been manufactured also from glass fiber or thermo-plastic materials, mainly polypropylene and PVDF (PVDF is more than ten times higher in material price than PP). Other materials are EP, PPN 7180 TV 20, copper stabilized PP etc.

Although glass fiber or glass fiber reinforced materials show a better chemical resistance than cast iron, corrosion still very easily occurs, especially when the surface has been damaged. In such cases a repair is not usually possible.

Comparing all these materials, it is shown that polypropylene is the better choice. The material properties, the chemical resistance, low weight and in comparison to the other material, lower production costs are self-explanatory.

Specific weight of: PP   0.92 g/cm3
                      Cast iron 7.85 g/cm3
                      PVDF   1.78 g/cm3

Material for Membrane Plates

The most important properties of materials used for membrane plates

Property	PP membrane material	EP membrane material	body plates
Density (g/cm3)	0.908	0.96	0.91
Melting point (°C)	160	165	165
Yield stress (N/mm3)	20	8.5	31
Flexural modules (N/mm2)	550	113	1150
Shore hardness	61 D	87 A	72 D
Average linear thermal expansion (mm/m/°C)	1.8×10^{-4}	1.8×10^{-4}	1.8×10^{-4}
Permanent Temp. permissible (°C)	100	100	100
FDA	yes	yes	yes

PP

Because of the chemical resistance of PP, its long term stability and its toleration of a wide temperature range plus other advantages, PP is replacing rubber, a material that was used before PP was known. In practice PP membranes can last two to five times as long as rubber membranes depending on product and operation.

Whereas rubber is very soft (Shore A hardness between 70—75) the PP material is comparatively hard (Shore D hardness 61).

The restriction of the flow rate path on the filtration surface of rubber membranes, caused by the deformation of the pips under feed and squeeze pressures is not seen with PP membranes. There are applications for example where rubber with its higher flexibility offers advantages over PP.

EP

An alternative to PP is EP. This thermoplastic material is rubber based but blended with PP. Because of the elasticity it can even be used under conditions below 0C.

Comparison between rubber and EP

Bending strength tests show that EP material achieves more than double the number of cycles than those of rubber.

Blending rubber with PP results in a reduced "aging" process. EP has a longer life span.

The hardness of EP membranes is approx. 87 Shore A. Rubber approx. 70 -75 Shore A. Due to the higher shore A hardness, there is hardly any deformation of the drainage pips thus the drainage between the pips is not restricted.

Despite its higher hardness the EP material's elongation is approx. 530 %, compared to rubber with approx. 450 %. This gives a higher security factor against membrane breakage.

Thickness of membranes in relation to temperature

PP	2 +1 mm - 0.2 mm	10 to 70°C
PP	3 + 1 mm - 0.2 mm	70 to 100°C
EP	4 + 1.2 mm	- 10 to 100°C
PVDF	2 + 1 mm - 0.2 mm	- 5 to 100°C
PVDF	3 + 1 mm - 0.2 mm	100 to 125° C

COPPER STABILIZED POLYPROPYLENE

A heavy metal stabilization prevents the oxidation which the heavy metals copper, cobalt and manganese can cause in polypropylene. The oxidation of PP produced by heavy metals (not heavy metal compounds) results in a cracking of the molecular chains and finally in a pulverization of the polypropylene.

The disadvantage of the heavy metal stabilized polypropylene is that the material does not correspond with the FDA regulations.

The heavy metal stabilization neither has a negative influence on the mechanical and thermal characteristics of the material nor on its general chemical resistance.

Copper stabilized PP should be used the moment there are the smallest traces of copper, cobalt and manganese, even when this is in the ppm range.

Only when thick filter elements like chamber filter plates and filter plate and frames are used metal concentrations of up to 20 ppm are acceptable and a copper stabilization is not necessary.

Polypropylene versus rubber (membrane plates)

Advantages of polypropylene
- better chemical resistance
- better temperature resistance up to 95 C
- better properties in sealing area pressure
- better hinge-properties, longer life-span

Disadvantages of polypropylene
- not flexible—cannot be stretched

Advantages of rubber
- greater flexibility—can be stretched

Disadvantages of rubber
- very soft and not pressure-resistant in sealing area
(deformation and consequently blocking of drainage depending on design)
- poor chemical and solvent resistance
- poor temperature-resistance
- short life-span depending on temperature and chemicals present in the slurry

Polypropylene versus cast iron

Cast iron plates were the standard filter elements for many decades. This began to change twenty years ago when polypropylene and other thermoplastic filter elements became available. Today, nearly two decades after the first polypropylene recessed chamber plates were molded, cast iron filter elements are very seldom found and the use of cast iron material for filter elements is normally restricted to high temperature applications where corrosion is not a problem. There are many reasons to change to polypropylene.

Cast Iron

Both spheroidal and ductile cast iron have been used as raw materials for filter elements, however, because of the low chemical resistance of the material to acids, both materials require some type of coating to protect them from oxidation and from chemical attack. Cleaning chemicals and other commercial cleaners attack cast iron.

Coatings

Because of the low resistance of cast iron to oxidation and chemical attack, coatings or linings are required to protect the integrity of the sealing areas and of the elements themselves. Three systems of coating have been used.

1. Rubber

Rubber compounds have been used to line or coat cast iron filter elements. The physical problem with rubber coatings is always the same. The high compression force exerted on the plate stack by the closure system of the filter causes the rubber to flow and deform. Eventually, the deformation in the sealing areas coupled with the natural aging properties of These cracks destroy the integrity of the sealing surfaces, cause leaking and eventually corrosion. The filter elements must be removed from the filter and completely re-lined. It is not unusual that re-lining of rubber covered filter elements has to be carried out during the first 3-5 years of service.

2. Thermoplastic linings

Development in lining cast iron filter elements has led to the use of various thermoplastic materials as lining or coating material. The material required to obtain a good bond with the cast iron is brittle and cracks both on the sealing surfaces under closure force, and on the drainage surfaces during filtration. Unnoticed cracks on the drainage surfaces lead to corrosion and bubbling of the thermoplastic material. Eventually, the elements require removal from the filter, surface stripping , and thermal re-lining. It is not unusual that thermoplastic re-lining of the cast iron filter elements is required every 6-8 years.

3. Liquid Set Coatings

Epoxy paints and other such materials have been used to protect cast iron filter elements from oxidation and corrosion. The life expectancy of a properly applied coating is normally 2-5 years. Recoating requires removal of the element from the filter, proper commercial sand blasting, and re-coating.

Cost of Re-lining/Re-coating

It is generally accepted that the cost of re-covering with any of the three types of linings or coatings exceed the cost of a new press pack of polypropylene filter elements.

Polypropylene

The compression molding process for polypropylene was developed in the late 1960's. Compression molding of thick pieces such as filter elements and their intended application requires a raw material with a fractional melt flow index (MFI) and a high E modules (elongation at break). The resin normally employed in compression molded recessed chamber plates is homopolymer PPH 2250 RAL 7032. This material exhibits excellent resistance to all inorganic acids and bases has a negligible natural aging time and is generally not subject to chemical attack or oxidation from most chemical compounds.

Weight

Density

The specific gravity of cast iron is approx. 7.85. The specific gravity of polypropylene is approximately 0.92

Chamber Weight

The weight of a typical cast iron filter element is approximately five times more than that of a typical polypropylene filter element designed for the same service.

Effect

The additional weight of a cast iron plate stack increase the cost of the supporting structure the filter press skeleton, the plate shifter, and the required overhead crane. Shipping and handling costs are substantially higher with cast iron elements than with polypropylene filter elements.

Cost

Filter Elements

The cost of a cast iron filter element is normally five times higher than that of a corresponding polypropylene element.

Effects on a Complete Filter Press

The complete cast iron filter press assembly normally will cost 50 % more than that of a filter press equipped with polypropylene filter elements.

General weight comparison based on a plate stack size 1200 x 1200 mm 100 plates and cake thickness 30 mm

Cast Iron	approx. 40 tons
PP	**5 tons**
Wood	8 tons
Rubber /Steel	20 tons

HANS G. VALERIUS PHD

COMPARISON OF DIFFERENT MATERIALS FOR FILTER ELEMENTS

Cast Iron	Glass Fibre	Polypropylene
High weight, difficult to handle. Requires high strength building foundation.	High weight	Low weight, can allow mobile filter press operation up to 200 m2 filtration area or more.
Low corrosion resistance even when lined (i.e. rubber) requires periodic relining at high costs	Low corrosion resistance due to unnoticed physical damage of surface.	Excellent chemical resistance against most aggressive media.
Membrane design only available with rubber membrane material. Natural aging and restricted chemical resistance of rubber reduces life expectancy.		Natural aging of PP nearly non-existent providing high life expectancy.
High mechanical strength can lead to plate breakage.	High mechanical strength but brittle can lead to plate breakage.	Limited mechanical strength minimizes plate breakage.
Portion of damaged elements may be recycled.	Cannot be recycled.	Up to 16 bar at ambient temperature Low probability for plate breakage due to highly elastic and resilient properties. Can be completely recycled.
Very expensive (100 %)	Expensive (60-70%)	Relatively inexpensive (20-25 %)

Size of filter elements

Filter elements from thermo-plastics are manufactured in sizes from 470 x 470 mm up to 2000 x 2000 mm. Chamber depths between 15 and 60 mm can be supplied. Usual sizes and designs are defined in the DIN 7129. Furthermore, elements can be supplied in inch dimensions.

Positions and diameters of filtrate and suspension channels differ between presses and depend upon filtration process. Consequently any existing filter press can be equipped with or replaced with molded elements from thermo-plastics. Even elements with top center or bottom center infeed can be refurbished.

Also depending on the product to be filtered, the size of the plate and the filtration process these filter elements can be equipped with stay bosses, i.e. supports that are integrated in the filtration area of each filter element.

Adapter plates may be used to convert from top center or bottom center to center or corner feed.

Production of the filter elements

There are three methods of production:
1. Granular molding is carried out by weighing the correct amount of PP pellets, placing them in the mold, heating for a period of time which will cause the PP to flow into a homogenous mass and then cooling. This method is normally only used for sheet material.
2. Frame/granular molding requires the pre-fabrication of a frame from sheet material. This is placed in the mold and the correct weight of granular PP is added. This overcomes problems of heat transfer related to difference in thickness of the sealing edge and the web of the plate. Molding then proceeds as described above.
3. Extrusion molding requires the use of an extruder in which the PP pellets are heated to a fluid mass and then extruded. The correct weight is transferred to the pre-heated mold and the process completed. This method gives higher production rates but cannot be used in all cases.

For all methods the PP must be allowed to cure, either by being supported over the whole surface area or by storing vertically.

This storage method is also required where finished products are to be kept for an extended period. They should also be protected from direct sunlight and kept at a temperature between 5—15 C.

The artificial aging process of PP can be achieved by thermosetting. Since PP has the tendency to work after production, changes in dimensions can be avoided by thermosetting. A change of length of 10—15 mm based on a plate size 1200 x 1200 mm can occur within a period of 3-4 weeks, due to creep. Temperature difference of 10 C calculated on a PP plate of 1000 x 1000 mm can result in a change in length of 1—5 mm.

4. Machining is carried out by normal practice and is used to square and plane and to drill filter plates for portage. In cases where non-standard filter plates are required they may be machined from slab material. Because of the thermal properties of PP it is only possible to recover a part of the scrap, for use of unstressed products.

5. Welding is used for fitting handles, plugging unwanted ports etc. The first method of welding is by pressing the two components on to a heated plate until the material flows. The welding "mirror " is then removed and the two parts pressed together while the material cools. The second method is by friction or " spin " welding and is used for closing unwanted ports. An oversize rod of PP is placed in a drill chuck, the drill is started and the rotating PP rod is offered up to the hole. The friction heats and flows the material and when the plug is in to the required depth the drill is stopped and held still to allow cooling, the drill chuck is released and the end of the plug is cut off.

Membranes are welded to the body plate to construct the one-piece version in a manner similar to the first method described. The difference being only that the welding mirrors are located to weld in the specific areas required.

Application of the different filter element types

Because of the different design of the various types of filter elements, important differences occur with their application.

With filtration on filter presses, the slurry feed and the build up of cake are done in an uninterrupted sequence.

With elements of fixed chamber depth, like filter plates and frames and recessed chamber plates, the dry solid content is dependent on the amount of dry solids injected into the chamber. A change of the characteristics of a particular suspension makes it necessary with these elements to re-asses the chamber depth and volume and also possibly the feed pump of the filter press.

Contrary to this, membrane elements have the capability of decreasing the chamber depth during the filtration. By applying air or water behind the flexible membranes they are inflated within the chamber. Increasing the pressure compresses the cake inside the chamber and squeezes liquid from it.

With this system, the dry solid content is now dependent upon time and pressure during squeezing. However, the structure of the solids in the chambers do influence the result during squeezing also. Comparison shows that dewatering with membrane filter elements is more effective than with standard recessed chamber plates. Usually, with membrane filter elements higher dry solid contents in the filter cake are achieved than with recessed chamber plates or with plates and frames.

This can be explained due to the fact that with recessed chamber plates only the feed pressure is responsible for compacting and dewatering of the cake. The feed pressure must dewater the cake from the feed hole tangentially by displaying liquid with solids. This is very ineffective and usually causes high variations in dry solid contents with a minimum effect around the feed hole. Time is another aspect when it comes to dewatering with recessed chamber plates. However, with membrane plates the pressure works normally perpendicular to the filter area. This effective direction of force compresses all sections of the filter cake equally and due to this leads to a higher dry solid content.

An increase of the overall throughput of the press is achieved when membranes are used.

To obtain a high dry solid content with recessed chamber plates the feed pump must keep on stream until all the chambers are filled with solids and no more suspension can be fed in. This process also shows that due to increasing flow resistance in the cake, the feed flow decreases dramatically during the filtration. In the last stage of filtration it can take a long time to increase the amount of solids in the chamber only slightly. With membrane plates, however the feeding process can be stopped before the chambers are totally filled. This reduces the amount of solids in the chamber and shortens the total cycle time. The result is a higher throughput.

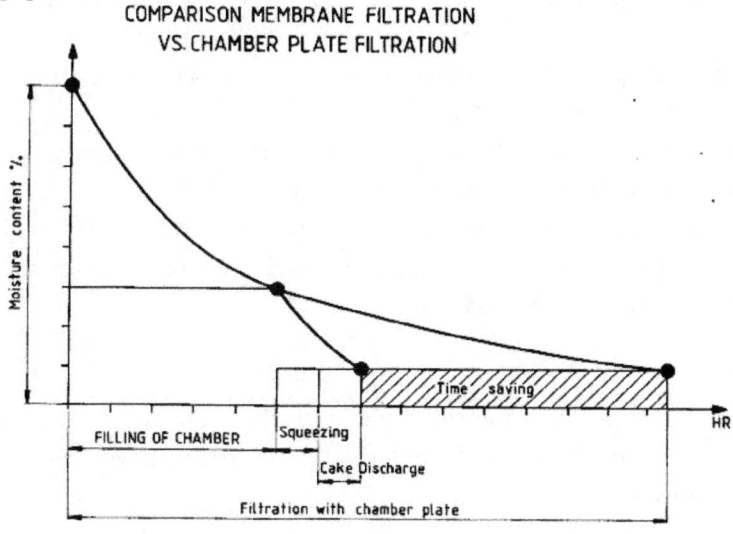

Cake washing

In many applications it is necessary to wash the cake after filtration. The aim of such a cake washing is either to replace the remaining liquid in the cake with an alternate liquid or to reduce the concentration of soluble chemicals inside the particles that form the cake.

All filter elements used with filter presses provide the opportunity for cake washing. Generally two different methods determined by the flow path of the wash liquid are possible.

1. Simple Wash

With the simple wash, the wash liquid is fed to the chamber by the suspension channel and runs through the cake like the filtrate.

The simple wash system pre-supposes that there are cakes formed on each cloth in the chamber and that semi-liquid interface remains, through which the wash water can pass into the cake. It is very difficult to control the cake build-up and the form of the interface and as a fully packed cake cannot be made the system is only marginally successful. Additionally there is the probability that, if wash water feed pressure is lost the cake will collapse into the bottom of the chamber, in this case a complete production batch can be lost. Although not very effective the simple wash system has an important advantage:

Simple wash can be carried out with any existing filter elements without modifying them. Therefore, a step with simple washing can always be added to any existing filtration process being performed with filter presses. After finishing a simple wash, the problem exists that a great deal of wash liquid remains in the filter cakes that dramatically increases the moisture content.

Therefore, the use of a simple wash with recessed chamber plates or filter plates and frames is not recommended.

Again membrane filter elements can remove this remaining wash liquid from the cake during squeezing and therefore achieve to a high dry solid content in the final cake.

2. Thorough wash

The second washing principle, the thorough wash, can, unlike simple wash, only be applied when the filter elements have been designed for this purpose.

LIQUID SOLID SEPARATION WITH FILTER PRESSES

Adjacent elements of a filter element pack are distinguished between pressure plates and wash plates. The difference is that the two elements have alternating filtrate outlets and therefore the filter element package furnishes two separate filtrate systems.

During filtration, both filtrate systems are connected externally and directed to a drain. For washing, these systems are separated. The wash liquid is fed through the filtrate system of the wash plates, spreads behind the filter cloth and flows through the cake due to the wash feed pressure. On the reverse side of the cake, the spent wash liquid leaves the element by the filtrate system of the pressure plates. Contrary to simple wash, all the chambers can be filled with cake during filtration due to this flow principle. Therefore, the chamber volume is fully utilized. Due to this principle of thorough wash, it is the preferred method of cake washing.

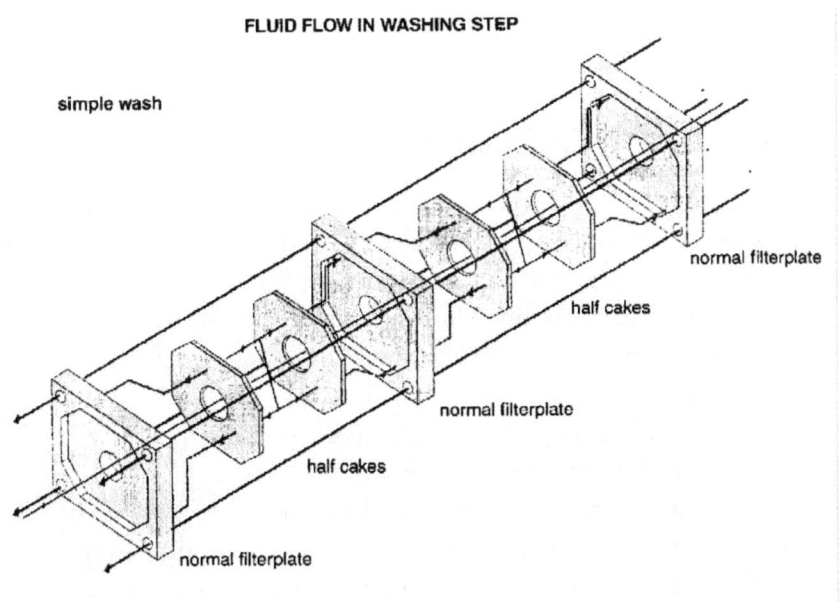

FLUID FLOW IN WASHING STEP

FLUID FLOW IN WASHING STEP

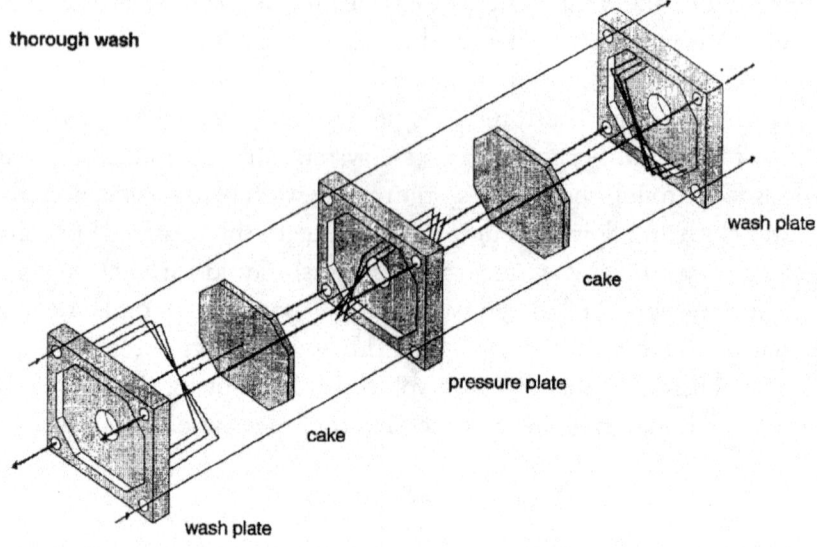

3. Main factors for cake washing

The wash liquid is forced through the cake by wash feed pressure. But the force for washing itself is determined by other factors and wash feed pressure usually is of either significance.

Furthermore, the main factors effecting cake wash are the chemical attributes of the participating substances since they preset the microscopic steps during washing.

When looking only at one particle of the cake, it can be seen that washing is marked by three steps.

First, the substance to be washed out must diffuse to the surface of the particle. The moving force for diffusion is only determined by the difference in concentration from the surface to the inner parts of the particle. Velocity of diffusion can only be increased by reducing the surface concentration.

From the surface of the particle, the substance must move to the wash liquid. To overlook the difference in concentration between the wash liquid and the particle surface is to ignore the single most responsible factor.

The velocity of moving over from the surface to the wash liquid can only be increased by increasing the surface area or increasing the difference in concentration.

Finally the substance to be washed out must be removed by the wash liquid.

This step is mainly preset by the solubility of the substances in the wash liquid and its flow velocity. Only this step can be increased by faster flow of wash liquid.

Increasing temperature can usually speed up each step during washing as the molecular movability of the substances to be washed out improves.

4. Border conditions for washing

The basic requirement for a homogeneous and liquid saving washing is a homogeneous flow through the cake.

The most important requirement for this is an even build up of cake. Additionally, cake washing is made complicated by a phenomenon called shrinkage cracks. Due to washing out of soluble substances in the cake and the resulting loss in volume, cracks occur in the cake which permits by passing of the wash liquid.

The borders of the filter chambers may also cause by-passing. In both cases wash liquid cannot pass through the cake homogeneously.

With thorough wash, the use of membrane filter elements provides a big advantage. By squeezing the cake with pressure slightly higher than the feed pressure of the wash liquid, the cake is permanently compressed during the wash cycle.

Due to this, occurrence of shrinkage cracks and by-passing in the flow of wash liquid is minimized or avoided. Also, the compression of the cake results in a more homogeneous permeability and therefore a more homogeneous wash.

Applying membrane plates means less wash liquid and shorter cycle duration.

5. Control and Regulation of the wash cycle

Regulation of the wash cycle is based mainly on the monitoring of the chemical and physical attributes of the wash liquid leaving the press.

Usually for regulation, pH, electrical conductivity, or turbidity of the wash liquid is used. If these characteristics, easy to monitor, are not sufficient for control and regulation during the wash cycle, periodically a chemical analysis in a laboratory should be made.

Furthermore, it has to be taken into consideration that when monitoring these values, only the characteristics of the wash liquid leaving the can be determined. From these values the content of the substances in the wash liquid can be calculated but this does not give a specific condition of the washed cake.

A precise control of the wash results is only possible with a continuous mass balance of the substances washed out during the washing cycle, the cost of which makes this impractical.

Therefore, based on the solids fed during filtration the amount of substance to be washed out must be estimated.

Monetary Aspects

The decision to use membranes is always combined with a high investment. Thus it is important to know the advantages that are offered with the membrane system and to estimate the return on investment.

Savings can be of different emphasis and depend on the product and the filtration process.

A precision study of where the advantages are when applying membrane filter plates should be done.

PP	2 +1	mm	10 to 70°C
	- 0.2	mm	
PP	3 + 1	mm	70 to 100°C
	- 0.2	mm	
EP	4 + 1.2	mm	- 10 to 100°C
PVDF	2 + 1	mm	- 5 to 100°C
	- 0.2	mm	
PVDF	3 + 1	mm	100 to 125° C
	- 0.2	mm	

Waste Water Treatment

For example, by decreasing the cake volume and weight of the material to be landfilled or incinerated in case of waste water treatment with membrane plates, a substantial monetary saving will be achieved. The dry solid content of the cake is increased and the moisture content consequently reduced. This automatically effects to the thermal energy costs. Due to the reduction of moisture content in the cake, less energy is required to incinerate the waste i.e. less water to evaporate, less B.T.U. (British Thermal Unit) is required.

This applies not only to waste water, but also to dyestuffs.

Labor costs

With membranes the overall cycle time is reduced and the volume handled is increased.(assuming that plate and frames or chamber plates were previously used and are now replaced by membranes). Due to the squeezing facility of membranes and to the fact that filtration can be stopped when the filter press is filled with the sludge to 50—75% of normal cake capacity (depending on the process) the cycle time can be reduced, thus the same output is achieved in less time. Where it is possible to utilize the additional available space in the filter press more plates can be inserted and consequently the output is higher than before. Again, savings which indirectly reduce labor costs.

More obvious are the savings in labor costs when the filter cake is discharged. Due to drier cake, the cake usually drops off easily when the press is opened. No extra time is required to scrape off the filter cake, which in some cases might stick onto the filter cloth and is hard to remove. The next step is the preparation of the filter press for the next batch.

Wash water costs

There are applications where the mother liquid has to be replaced by a different liquid. Or in other cases soluble salts or other liquids have to be removed from the cake by rinsing it with water or any other liquid. This kind of applications can be found in the dye and pigment industry, in the chemical and in the mining industry.membranes contribute to the reduction of volume and/or time required to achieve the above aim.

The contribution might vary with the application, however the aim is achieved by:

Applying a low pressure squeeze before and during the wash cycle so that the cake will be less susceptible to channeling.

Applying a low pressure squeeze before and during the wash

cycle, much of the mother liquor in the cake that has to be washed is removed. Thus less wash liquor is required to displace the mother liquor.

Applying the final squeeze wash liquor is removed.

Wash liquor disposal costs

The wash liquor can either be a product of the process itself or it can be considered as waste liquid. Regardless of which type of the two the wash liquid belongs, there are savings. In case of waste liquor, the savings by reduction will vary, i.e. can it be sent to the municipal sewer or has to be treated in an on-site treatment plant. In other cases, assuming that the chemicals are hazardous, special handling is required. Savings can be calculated by knowing the change relative to volume and the reduction in volume resulting from the membrane system.

Increasing capacity

By applying membranes and replacing existing chamber plates or plates and frames, the size of the filter press can remain while the output is increased. This can be achieved in two ways. First, the cycle time is shortened, which means more cycles per day; second the total filtration volume and area is increased which means another increase. By avoiding respectively by not changing the size of the filter press, there are savings which are achieved automatically, such as no additional floor space required, no additional manpower etc.

Higher yield

In simple words, higher yields are obtained by the squeeze system in a process where the higher value is the filtrate i.e. in the palm oil industry—the olein.

Non—Monetary Aspects

The non-monetary advantages of applying membrane systems can be found in:

Higher quality of the end product, i.e. greater advantages in marketing the product in comparison with the competition.

Better basis to meet environmental laws and regulations.

Flexibility in the filtration process.

Squeeze system

Membrane filter elements provide the possibility to compress the cake and increase the dry solid content in the cake by applying a pressure medium behind the membranes.

The squeeze media can be conveyed to the filter plates by individual connections at each filter plate or through a centralized system through the complete filter press pack.

1. Choosing the squeezing medium

In the past compressed air was mainly used as the squeeze medium. Recent safety regulations have resulted in application of water or product instead of air. Above 7 bar air should not be used.

Energy is stored by feeding a squeeze medium behind the membranes inside the filter press. The amount of stored energy basically depends on the compressibility of the applied squeeze medium and the amount of pressure. Malfunction of the filter press closing mechanism can result in the sudden and uncontrolled release of this energy resulting in uncontrolled expansion of the filter element package.

The high compressibility of air requires large amounts of energy for compression. When using water as a squeeze medium, its low compressibility causes storage of only 2 % of the energy as compared with using air at the same pressure and volume.

Therefore the safety threat posed by the squeeze media in the event of a malfunction of the closing mechanism as well as the energy required for pressure increase can largely be reduced by using water. Because of this, squeeze systems based on water or product has lower operating costs than systems with air.

Furthermore, the investment costs of a squeeze system also depend largely on the pressure medium chosen. The high compressibility of air requires expensive compressors with high volume supply capabilities where normal centrifugal or screw pumps are sufficient when using water.

Unlike air, there are few points to be taken into consideration when using water:

While compressed air leaves the filter element package naturally at the end of squeezing, using water requires a discharge pump or a significant difference in height.

Furthermore, the low compressibility of water may cause rapid loss of pressure with even small leaks in the system.

Technical and economic aspects govern the choice of the squeeze media. The question whether to use individual or central supply of the squeeze media also has to be considered.

Individual air or water supply as squeeze media, i.e. each filter element has one hose connection results in an arrangement of hoses that have to take into consideration the length of each hose in accordance with the distance the filter elements have to move when the cake is being discharged.

The advantage that is offered by the individual squeeze system is found in the easy detection of leakages within the squeeze system.

2. Designing a squeeze system

In the attached flow chart a water based squeeze system is illustrated.

The vessel is used to store the squeeze water, so that the same water is used repeatedly and only losses from evaporation require additional water from time to time. The squeeze pump takes water out of the vessel and feeds it to the distributor pipe next to the filter press. Each membrane filter plate is connected to this distributor pipe with flexible hoses that allow movement of the plates for cake discharge.

3. Safety devices

Even with a squeeze system based on water, safety interlocks must be incorporated into the squeeze system. In the first place operation of the squeeze pump should only be possible when the closing force of the press is sufficiently high. The second safety device must limit the maximum internal squeeze pressure produced by the pump. To avoid the pump exceeding the required maximum, a relief valve with a dump line back to the vessel is recommended. Moreover, the squeeze pump should be protected against running dry and supplied with external cooling that avoids over-heating with long runs at low rpm.

A third interlock must ensure that the press cannot be opened while squeeze pressure exceeds 0.2 bar.

To fully exploit the advantages of the low compressibility of water, the squeeze system must be totally evacuated of air on startup. Incomplete evacuation of air can cause problems with evacuating the water, especially when the flexible hoses are fixed to the top of the plates.

COMPARISON OF WATER AND AIR USED AS PRESSURE MEDIUM

Compressed Air	Water
Very high amount of energy inside the system with operating pressure (50 times higher than with water), coupled with slow pressure release poses a threat.	Low amount of energy inside the system with operating pressure, coupled with fast pressure release poses no threat.
Energy requirement is very high and causes high operating costs.	Low operating costs due to low energy requirement.
Delivered pressure must be higher than required for operation.	Delivered pressure is same as pressure in operation.
Squeeze systems require high investment costs, due to the requirement for compressors with high air delivery rates.	Squeeze system require only low investment costs.
Small leaks in the squeeze system can be tolerated.	Even small leaks in the squeeze system may cause problems.
The pressure medium releases by itself.	The release of the pressure medium requires pumps or gravitational difference.

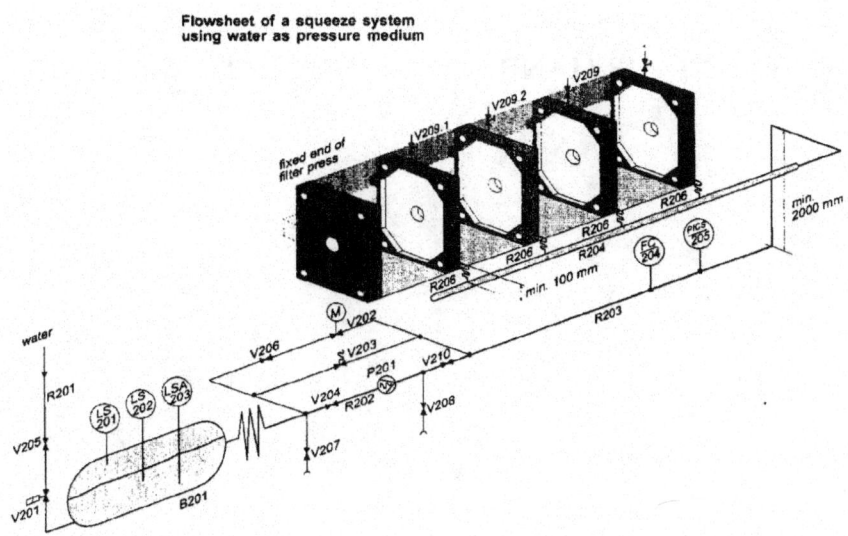

Flowsheet of a squeeze system using water as pressure medium

LEGEND

B 201	water storage vessel
R 201	city water connection
R 202	suction pipe
R 203	feed pipe
R 204	distributor pipe
R 206	flexible connection hose
P 201	eccentric screw pump
V 201	solenoid valve
V 202	auxiliary discharge motor valve
V 203	safety valve
V 204	auxiliary valve
V 205	auxiliary valve
V 206	auxiliary valve
V 207	air ventilation and drainage valve
V 208	air ventilation and drainage valve
V 209	air ventilation and drainage valve
LS 201	maximum level switch
LS 202	minimum level switch
LSA 203	low water alarm and protection switch
FC 204	flow indicator
PICS 205	pressure sensor and safety switch

MEMBRANE AIR MANIFOLD—FEED, DRAIN, WASH, AIR BLOW PIPING

LIQUID SOLID SEPARATION WITH FILTER PRESSES

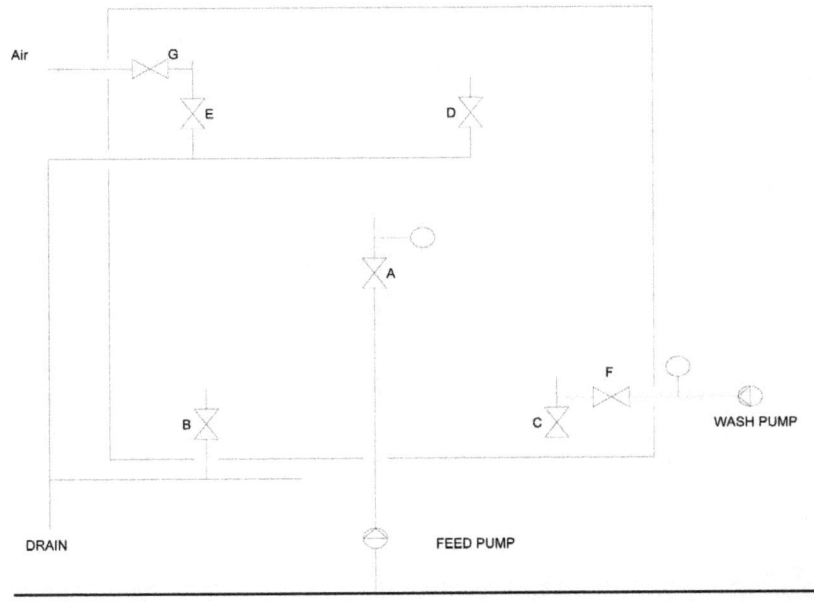

CHART	VALVE						
	A	B	C	D	E	F	G
1. Feed	O	X	X	O	O	X	X
2. Pack	O	O	O	O	O	X	X
3. Pre-Squeeze	X	O	O	O	O	X	X
4. Wash	X	X	X	X	O	O	X
5. Squeeze	X	O	O	O	O	X	X
6. Air blow cake	X	X	O	X	X	X	O
7. Vent Membranes	X	O	O	O	O	X	X
8. Air blow Lines	X	O	O	X	X	X	O

NOTE: 6 +8 may not be required in which case valves F & G will be omitted.

Schematic—Membrane air inflation system for pre-squeeze and squeeze operation.

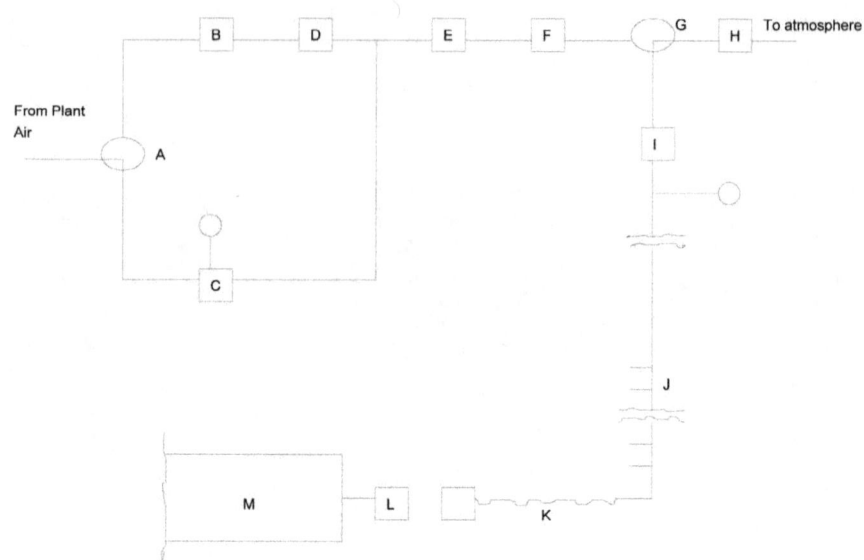

Legend

A. 3-way valve
B. Squeeze pressure regulator
C. Pre-squeeze pressure regulator
D. Pressure amplifier
E. Valve-Interlocked to open only when press is at full closing load.
F. Orifice plate
G. 3-way valve
H. Muffler
I. Pressure Switch-Interlock to prevent press opening while membrane air pressure exceed 3 p.s.i.
J. Manifold
K. 1/2" hose complete with self-sealing quick-disconnect 1/2 coupling (female) (1 per membrane)
L. Open quick-disconnect 1/2 - coupling (male) (1 per membrane).
M. Membrane plate

How to size a compressor

All the formulas are based on a final cake thickness of 75 % of the chamber depth, i.e.

40 mm cake thickness with a 30 mm cake.

    C   total chamber volume (ltr)
    B   maximum squeeze pressure (bar)
    T   time to reach B (from squeeze diagram)
    D   chamber depth (mm)

Air compressor rating

ltr/min free air delivered = $0.25 \, C \times B : T$

LIQUID SOLID SEPARATION WITH FILTER PRESSES

Example: C = 3000 ltr
B = 7 bar
T = 3 min

0.25 x 3000 x 7 : 3 = 2.500 ltr/min free air

The pressure maximum should be at least 10 % higher than the maximum the press requires. After the compressor a regulating valve has to be installed to reduce down to the maximum squeeze pressure. To control the rate at which squeeze pressure is increased, an orifice plate is installed between the regulator and the membranes.

Size of orifice (diameter mm):   For temperatures above 10 °C (5.49 C : D : B)
For temperatures below 10 °C (2.75 C : D : B)

Example C = 3000 ltr
D = 40 mm
B = 7 bar

2.75 x 3000 : 40 : 7 = 5.43 mm diameter

Squeeze water pump ltr/min   0.25 x C : T

Example C = 3000 ltr
T = 3 min

0.25 x 3000 : 3 = 250 ltr/min

To determine the size of the water tank as a good guidance the same capacity as the press should be taken.

Air blowing

Fixed chamber presses often produce a cake that is not as dry as required and to resolve this, air is passed through the cake but in the reverse direction to the wash. This is usually carried out after the wash. Membrane plates usually only require air blowing when solids are very granular and cannot be compacted beyond at certain point. Air blowing is carried with the squeeze pressure applied so that there are no voids in the chamber which would allow air to bypass the cake.

Core wash / blow

Because membrane plates apply force only over the filtration area a wet core is sometimes left in the feed eye, which would fall into the cake. To overcome this it is possible to inject water or air, or both to transfer this wet material back to tank or to waste. This is done by a connection in the tail end plate and an additional valve line bypassing the feed valve and pump. As with air blowing squeeze must be maintained to prevent the wet core running down the face of the cake.

This system is rarely needed with fixed volume presses as the feed eye is packed.

Venting

At this point membrane squeeze would be released.

Line blow

Since a certain amount of liquid is always retained on the surface of the plates and in the portage it is sometimes desirable to use available air to remove this. Normal practice would be to close the top drain valves, apply air at a top inlet and allow air to carry surplus liquor out through the bottom drains.

Sterilization of membrane plates

Membranes can be sterilized with steam up to 120 C° at 1.9 bar for 30 minutes feed time, squeezing after 10 minutes pause for 3 minutes at 0.5 bar. This sterilization should be done max.once a week.

Accessories

A filter element requires a number of additional parts before it can be installed in a press. It also requires identification.

1. Cloth pins

Cloth pins are fixed on top of the filter element to serve as point of fixture for the filter cloth. The number of cloth pins depend upon the size of the plate.

These cloth pins are of different color:
black—for head and end plate
red—for pressure plate
blue—for wash plate

Other arrangements are also possible, i.e. cloth pins can be made from stainless steel. Marking can be done on the side of the filter plate.

2. Production and quality control stamps

Normally the filter element bears the month and year of production on the side as well as the quality control stamp.

3. Handles

Depending on the filter press construction filter elements have to be equipped with handles to support the filter element on the side bars of the filter press. Exceptions are the filter elements that are for overhead suspension presses, here are no handles necessary. Basically there are the standard handles and the special types in accordance with the filter press manufacturers design.

4. Discharge valves / spigots

Discharge valves or spigots are available in single outlet (permanently open), two-way and four-way types (see under " Technical clarification ").

5. Filter cloth locking devices

Dependent on the design of the filter element center or corner feed filter cloth locking devices are necessary where barrel neck filter cloths are not used in order to fix the filter cloth in the feed port. Filter cloth locking devices are available in stainless steel and PP as well as other materials.

Technical clarification

Details essential to know

Whether a complete filter press or only filter elements or filter cloth is required, some details have to be checked prior to ordering. The press manufacturer, plate manufacturer and cloth supplier all have a need to know the details of the process in terms of the filterability, material, compatibility, temperature, pressure etc.

They also need information on the size, type and location of feed ports, drain ports, handles, spigots and cloth pins as well as the chamber depth. Temperature, pressure and material compatibility are particularly important in making a selection between PP, EP, PVDF or other materials such as PPN 7180 TV 20. This is also critical in choosing between homo- or copolymer and even copper stabilization.

Similar information is also required by the cloth supplier who also need details of the filter plate design in order to exactly tailor the cloth to the plate, a poorly fitted cloth can create severe problems in the operation of the press.

The first requirement is a knowledge of the components of the press, the terminology, the fact that all dimensions are taken from the horizontal and vertical center lines of the press end and that all descriptions are based on viewing the press from the outside of the feed (piping) end.

Selection of the type of filter element can be made by either the filter press manufacturer or the filter element manufacturer, based on existing experience with the filtration of a specific product or based on test filtration. These are strongly recommended in cases where latest technology has not yet been applied.

The filtration tests will show which type of filter plate should be used. The desired output determines the number of plates, the volume, filtration area, design and cake thickness of the plates,

There is a wide variety of aspects.
1. Installation of complete filtration system
2. Replacement of existing filter elements by same type of filter elements
3. Replacement of filter plates and frames by chamber plates
4. Replacement of chamber plates by membrane plates
5. Conversion of filter elements of different material into another filter element material.

In many cases, it is advisable to investigate whether product quality, cake dryness, output etc., can be improved by applying membrane plates. If no data is available, the logical step is the test filtration. Since there is no written law in the solid-liquid separation with filter elements, filtration tests are a must considering that each product is different and even within one product the parameters can vary. Test filtration is also the final basis for the decision on the right choice of filter cloth.

The description of the filtration objectives is the most important issue.

Operating filter elements

General Operating Instructions

1. An optimization of the operation of filter plates and frames,

chamber plates and membranes must be established in practice.
2. Filtration- and washing pressure depends on operating temperature. Limitations are shown in the attached diagrams.
3. The membrane squeeze pressure rise and temperature limitation are shown in the attached diagram.
4. For the membranes, squeezing at maximum pressure is generally limited to 15 minutes, however, there are meanwhile applications exceeding this limit. The plate manufacturer has to be consulted.
5. Shock temperatures must be avoided. A maximum difference of 50°C must not be exceeded.
6. The press closing system and the membrane squeezing system must be interlocked to ensure that:

The press cannot be opened while the membrane squeezing pressure is higher than 0,2 bar.

The squeeze pressure cannot be applied until full press closing force is applied.

7. When squeeze pressure is not being applied, the membranes must be vented.
8. The final cake thickness of all membrane plate packs after squeezing must not be less than 50% of the chamber depth.
9. The final cake thickness of all mixed membrane—chamber plate packs after squeezing must not be less than 75% of the chamber depth.
10. Differential pressures have to be avoided.
11. The fit of the squeeze grommets and filter cloths must be checked regularly. The filter cloth must not restrict the filtrate outlet.
12. Feed and drainage ports as well as stay bosses and gasket faces should be kept free from any deposits.

The above are guidelines only. It is strongly recommended to refer to the operation manual. Operators should be trained accordingly and the filtration plant should have adequate recorders installed.

How to re-gasket filter plates

As mentioned earlier the non-gasketed filter plate version provides a plate-to-plate seal where leakages might occur during operation.

The gasketed version is nearly leak-free which is given by the gaskets around the filter plates. The gaskets protrude approx. 0.030 to 0.060 " and guarantee the seal between the filter plates. O-ring type gaskets are installed not only around the sealing surface of the filter plate but also around the outlet eyes.

For installation the gasket has to be cut squarely at the end. It is inserted at the bottom center of the filter plate by using a mallet. It has to be considered that the gasket is cut up to 1" longer than required. The end are connected by using glue and the excess of 1 " is pressed into the groove. Gaskets have the tendency to creep; thus it is necessary to work with an excess length of max. 1 ".

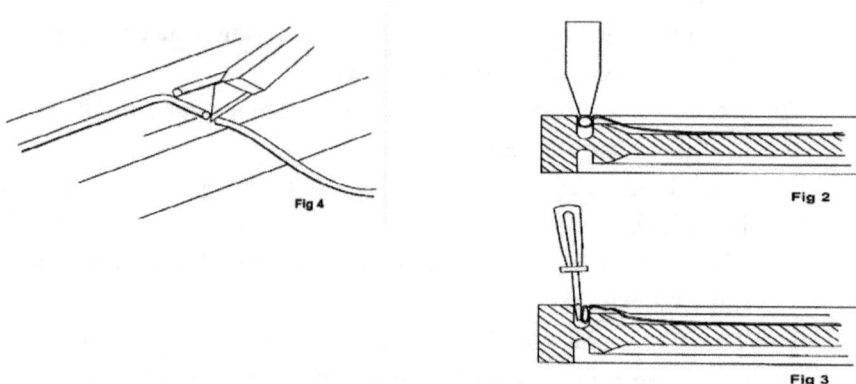

Fig 4

Fig 2

Fig 3

The life of the gaskets will mainly depend on the length of the filtration time, temperature, closing force etc. At the beginning of the filtration new gaskets might pull out of the grooves when opening the filter press and separating the filter plates. The application of a silicone spray will help for the time being and after a few filtrations have been done this minor problem is solved automatically since the product creates a film on the surface after several cycles.

How to fix filter cloths

There are basically two versions:
1. non-gasketed
2. gasketed

Cloth for the non-gasketed version are hung over the filter plate from the top to bottom, held by the eyelets which match with the cloth pins on top of the plate. The seal is provided by the plates pressed together. Some sort of leakage during operation can occur which is to be considered as normal for this kind of filter cloth installation. The gasketed version reduces the leakage and is nearly leak-free due to the gasketed area around the chamber and the filtration ports. The gasketed cloth is connected at the end with a high-density cord which is sewn around the ends and pressed into the groove on the plate around the filtration area. The grooves on the filter plate are approx. 3/8 " wide and 3/8 " deep. The cord diameter normally is of same size, however in some cases will depend on the type of filter cloth and its thickness. A screwdriver with an appropriate blade is inserted into the groove to move out a small section of the cloth, the remaining inserted cloth is simply pulled out. New filter cloth can be inserted after the groove is cleaned (possibly accumulated solids).

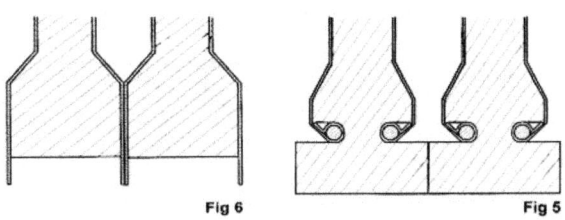

Fig 6   Fig 5

The removal of filter cloth (non-gasketed) is as easy. The ties at the side of the filter plates (if used for the installation) are simply cut and the cloth is pulled down by removing the eyelets from the cloth pins.

Washing of filter cloths in situ

Washing filter cloths with a jet of water under high pressure may be insufficient if the slurry has a tendency to crystallize. In such a case the filter-press will be filtered with a washing device which operates when the filter-press is closed, by circulating inside the filter-press solution (acid for instance) capable of dissolving the product encrusted in the filter cloths.

The solution is prepared in a tank, and delivered to the closed filter-press by a centrifugal pump, recirculating to the tank until a satisfactory cleansing is obtained (for instance until the acid solution is neutralized). The operation usually lasts 1 or 2 hours.

It can also be done by successive fillings and drainings.

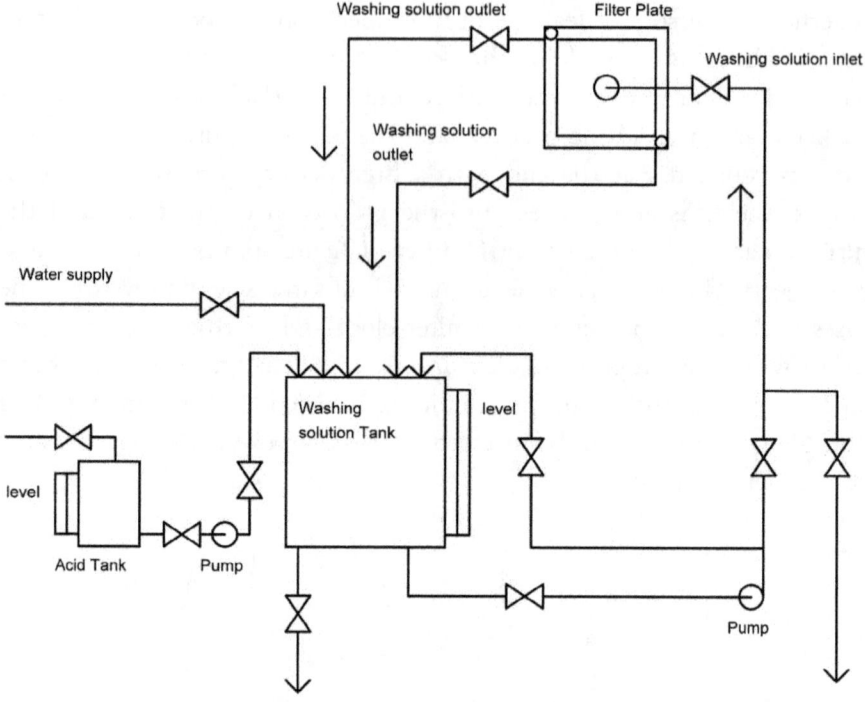

Washing of filter cloths without removing them on a closed filter-press.

Regarding washing of filter cloth also refer to "Filter Presses"

Trouble shooting

In many cases customers will state that the filter plates are faulty with little evidence to support the statement. The following subjects refer to general statements and the possible reasons:

Leakage from the plate edges

Insufficient closing load.
Fold in filter cloth
New filter cloth
Build up of cake on sealing edges
Plates bent so that sealing faces do not lie flat on each other
Damaged sealing faces
Press over-packed (this can occur even when correct closing force is applied)
Cloth misaligned at corner eyes
Seals of connection / Adapter plates damaged
Seals of caulked and gasketed plates are damaged
Insufficient slurry
Incorrect cloth allows solids to pass
Damaged cloth allows solids to pass
Misalignment of cloth between corner eyes and chamber allows solids to pass

Wet cake

Insufficient pump pressure
Cloth blinded

Bent plates / thick / thin cakes

Improper filling
Plugged drain ports
Plugged feed ports
Blinded cloth
Drainage surface blocked
Mixture of different cloth types

Poor washing

Poorly formed cakes
Insufficient wash water
High or low wash rates
Plates installed incorrectly

Membranes

Hinge area white (normal because the molecules re-align under stress) Bulging membranes (often occurs with age or overstretching)

> Will not hold pressure :   check all connections first
>   grommets missing
>   cloth over grommets
>   split membrane

Tracing a leaking membrane

Grommeted squeeze systems make tracing a leaking membrane a lengthy process. Each plate has to be tested individually up to not more than 0.25 kg/cm2 by applying air through one grommet hole. Soapy water over the plate or complete submersion can help.

With the external squeezing system it is easy to remove the hoses from the plates. Then the press can be filled under pressure and liquor will come from the connection of the plate which is normally used for the squeezing media.

From the filter press manufacturer's point of view there are also a list of reasons for possible interruption which are identical with the ones indicated by the filter element manufacturer, however, some of those interruptions refer only to the filter press itself.

LIQUID SOLID SEPARATION WITH FILTER PRESSES

Unclear filtrate

Damaged filter cloth -   Check on same

Caulking out of grooves -   Replace caulking (caulked & gasketed version)

Filter cloth is pulled out of groves
(caulked & gasketed version)

A full cake was not built up -   To avoid that the cloth is pulled
before washing   out of the groves a proper
                                           cake built - up is necessary
                                           otherwise the cake cannot support
                                           the cloth.

The cord (O-ring) / or the filter cloth -   Change size of future cloth
is not properly sized

Follow—up for maintenance of filter presses

Daily

    Clean sealing areas of solids build—up
    For caulked & gasketed version clean area and replace gasket with signs of damage, i.e. cuts etc.
    Check on filter cloth damages, like holes etc.
    Check on any leakage.

Weekly

    Check on oil level in hydraulic reservoir
    Check on relief valve setting
    Wash filter cloth depending on process (with caustic or acid)

Monthly

    Clean oil filter
    Wash filter cloth depending on process (with caustic or acid)

Yearly

    Replace oil filter
    Replace oil in hydraulic system
    Replace gaskets if any

Out-of-balance pressures

Damage of filter elements in filter presses are in most of the cases due to so-called out-of-balance pressure which results in broken web and/or deformation. The signs of such a damage are very obvious. They are mainly caused by:

1. Different permeability of filter cloth or wrong selection of filter cloth, i.e. the filter cloth installed is not suitable at all for the filtration task.
2. Feed holes are blocked since they were not cleaned after the discharge of the cake after the filtration done before.
3. Feeding speed too low or too high.
4. Different cake thickness in the same filter press.
5. Not properly sewn filter cloth especially with corner feed designs and possible crystallization of the product during the feeding process. Another possibility is that also here for corner feed so-called barrel neck design is used instead of filter cloth locking device resulting in a very small space left open for the product to be fed into the individual chambers.

The results can be easily determined by thinner or thicker cakes all over the filter press and in extreme cases a higher cake thickness than actually the plate is designed for. In such a case the web already gave way and can be considered as damaged.

LIQUID SOLID SEPARATION WITH FILTER PRESSES

UDC 66.067.4-22 DEUTSCHE NORMEN

Filter presses DIN 7129

Recessed plates, frame

plates, frames, supporting bars

Main dimensions, tolerances

Dimensions in mm

1 Scope

This standard applies to
- square recessed plates for suspension from lateral round bars or rectangular bars
- square frame plates for suspension from lateral round bars or rectangular bars
- square frame for suspension from lateral round bars or rectangular bars
- rectangular recessed plates for suspension from lateral rectangular bars for filter presses

Additional statements are made concerning the supporting bars and cake thickness. The handles to be provided on the sides of the plates or frames for suspension from the supporting bars are not covered by the standard.

2 Recessed plates, frame plates, frames, supporting bars

2.1 Square recessed plate (QK) or frame plate (RP) or frame ® for suspension from round bars

Square recessed plate (QK) or frame plate (RP) or frame ® for suspension from rectangular bars

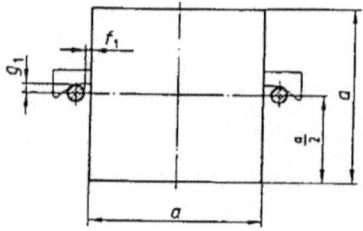
Fig. 1.

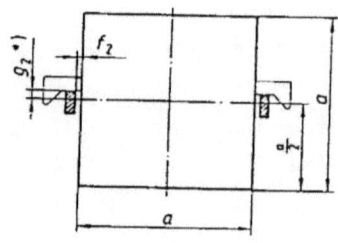
Fig. 2.

1) See explanatory notes

DIN 7129

2.2 Rectangular recessed plate (RK) for suspension from rectangular bars

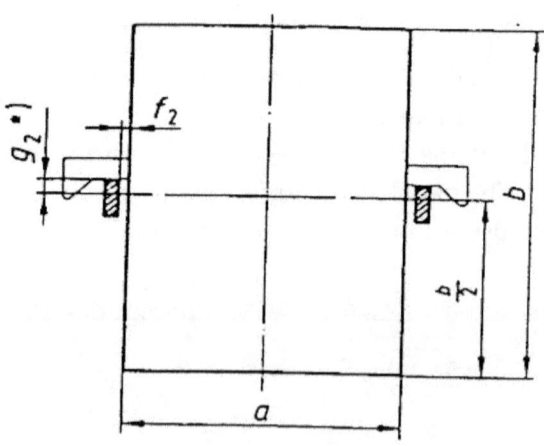

Fig. 3.

Table 2. Rectangular recessed plates and rectangular bars

Size of recessed plate (nominal size)	Rectangular bar Clearance between bar and recessed plate
a x b f2	min.
1200 X 1600	60
1300 X 1900	70
1500 X 2000	70

*) See Explanatory notes
2) Other bar profiles are possible
DIN 7129

Table 1. Square recessed plate or frame plate or frame and supporting bars

Size of recessed plate or frame plate recessed or frame (nominal size)	Round bar			Rectangular bar
	Clearance between bar and recessed plate or frame plate or frame	Half bar height		Clearance between bar and plate or frame plate or frame
a	f1	g1		f2 min.
300	10	17,5		20
500	20	30		20
630	20	40		30
800	20	50		35
1000	20	55		35
1200	25	60		40
1300	--	--		40
1450	25	62,5		50
1500	25	62,5		50
1800 1) see below	--	--		60
20000	--	--		70

1) This size was specified for possible future requirements.

DIN 7129
3 <u>Frame plates (RP), frames (R) and size of ducts</u>
Type A

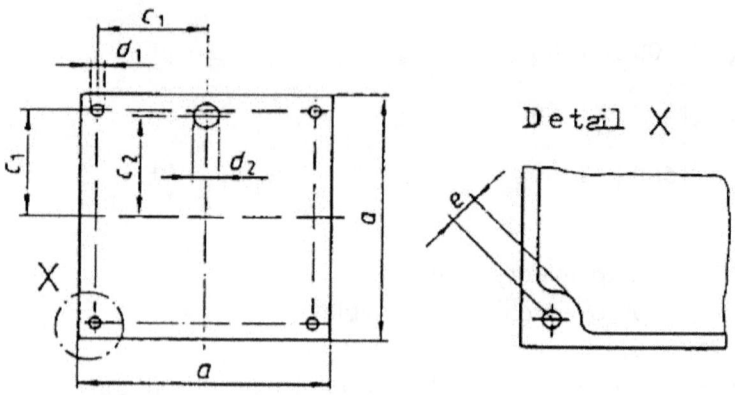

Fig. 4.

Designation of a frame (R), type A with a = 1200 mm, for a cake thickness m = 25 mm:

Frame DIN 7129-R-A 1200 x 25

DIN 7129
Table 3. Frame plate or frame

Size of. Frame plate or Frame (nominal size)	C1	C2	d1 = d2	
a			max.	min.
300	118	--	25	30
500	200	--	40	45
630	255	250	50	55
800	335	335	50	60
1000	430	425	50	65
1200	520	500	65	75
1300	565	550	80	80
1450	640	630	80	80
1500	665	--	80	80
1800 1) see below	800	--	100	95
2000	895	--	100	100

1) See Table 1

DIN 7129

4 Recessed plates, position and size of ducts

4. 1 Square recessed plates (QK)

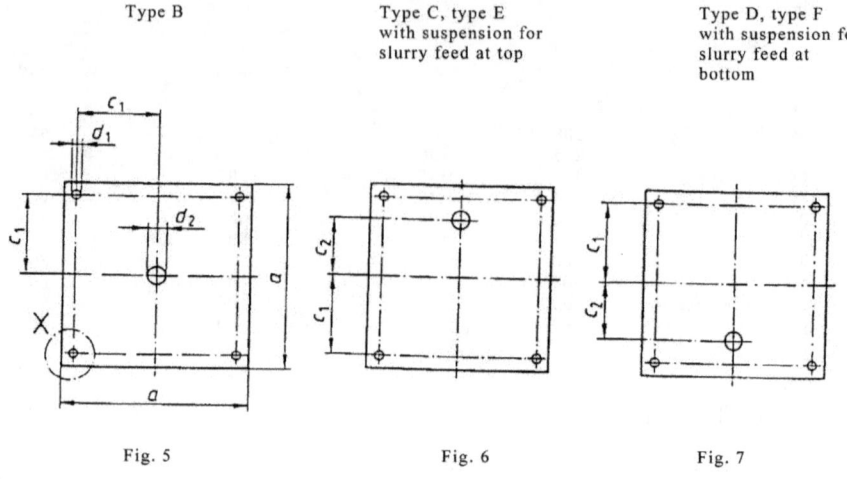

Type B

Type C, type E
with suspension for
slurry feed at top

Type D, type F
with suspension for
slurry feed at bottom

Fig. 5 Fig. 6 Fig. 7

Designation of a square recessed plate (QK), type C with a = 1000 mm, for a cake thickness m = 50 mm :

Recessed plate DIN 7129-QK-C 1000 x 50

LIQUID SOLID SEPARATION WITH FILTER PRESSES

DIN 7129
4.2 Rectangular recessed Plates (RK)

Type G

Type H
with suspension for
slurry feed at top

Type J
with suspension for
slurry feed at bottom
Other dimensions as type G

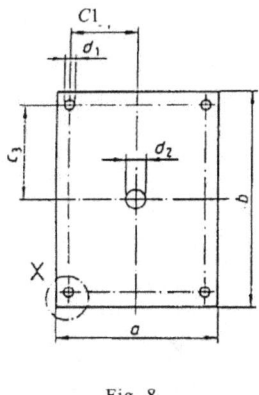
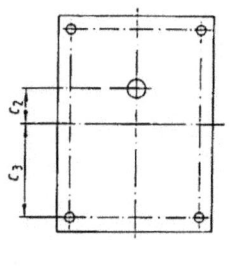
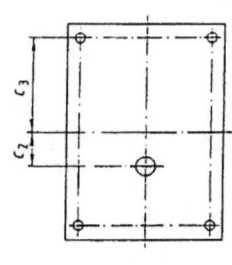

Fig. 8   Fig. 9   Fig. 10

Designation of a rectangular recessed plate (RK), type G with a = 1300 mm, for a cake thickness m = 40 mm:

Recessed plate DIN 7129-RK-G 1300 x 40

DIN 7129
Table 5. Rectangular recessed plates

Size of recessed plate (nominal size) a x b	C1	C3	C2 3) Type H and J max.	d1 max.	d2 max.	e min.
1200 x 1200	515	715	265	80	150	80
1300 x 1300	560	860	315	80	150	85
1500 x 2000	645	895	335	100	200	100

3) See Table 4

5 Cake thickness

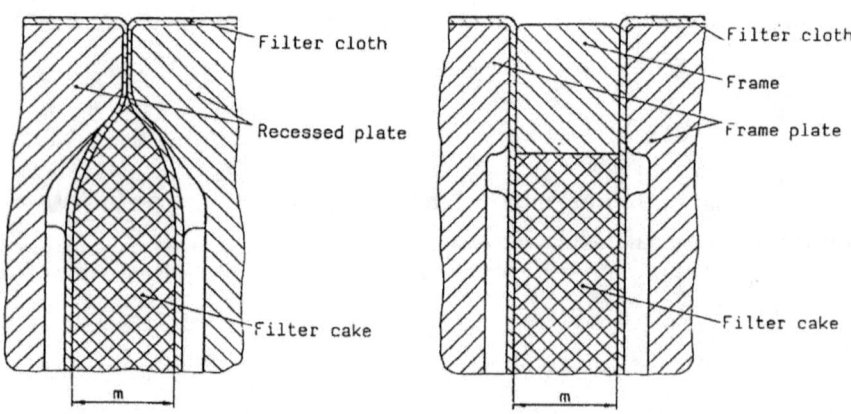

Fig. 11 Recessed plate   Fig. 12 Frame plate or frame

DIN 7129

The cake thickness (m) is the dimension, which the filter cake may assume between the filter cloths.

Table 6. Cake thickness

Cake thickness (nominal dimension)	15	20	25	32	40	50	63	80

6 Tolerances

6.1 Parallelism and flatness

The sum of the deviations from parallelism and flatness of the sealing face on the recessed plate or frame plate or frame shall not exceed 0.2 mm.

Note: Confirmed experience has been acquired with this deviation up to size a = 1500 mm of recessed plate or frame plate or frame. Special agreements shall be made if necessary for the larger sizes.

6.2 Thickness

The tolerance for the thickness h, which is selected by the manufacturer, of the recessed plate, frame plate or frame is o mm.

DIN 7129

Recessed plate   Frame plate   Frame

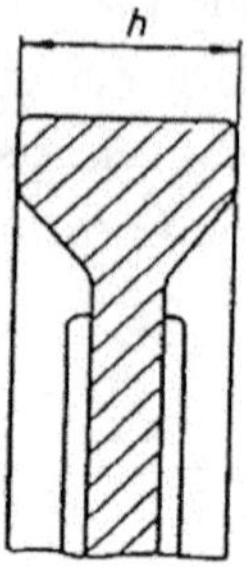

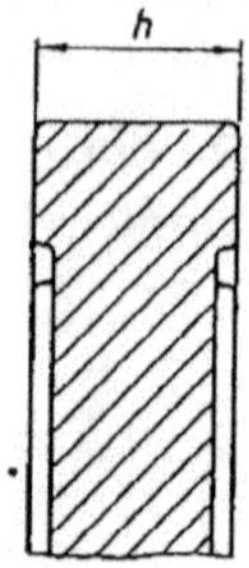

Fig. 13   Fig. 14   Fig. 15

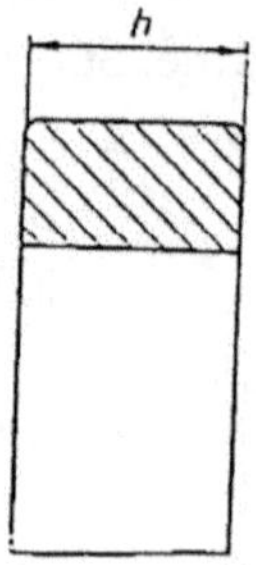

6.3 General tolerances
General tolerances DIN 7168—coarse.

7 <u>Material</u>
As agreed.

Further standards
DIN 7168 Part 1 General tolerances (dimensional variations); length and angular dimensions
    DIN 7129

<u>Explanatory notes</u>

DIN 7129 was revised to meet the requirement of the filter press manufacturers for standardisation of the distances between plates and bars. During the negotiations it emerged that there was agreement concerning substantial extension of the standard, so that this standard is supplemented by statements concerning rectangular shapes, position and size of the ducts, cake thickness and plate tolerances compared to the May 1971 edition.

As there was evidently no longer any interest in the circular plate shape, it was omitted. It was not possible to retain the hitherto small number of sizes for square plates. As a result of different works standards individual operators were equipped almost exclusively with plate size

1450 or 1500. As filter presses are long-lived and highly expensive commodities, omission of one or the other size would have resulted in an unnecessary expensive investment as a result of conversion. The plate sizes 1300 and 1300 x 1900 were included predominantly at the request of firms in neighbouring countries.

The diameters d1 and d2 are maximum dimensions and are designed to delimit the space requirement and position of the holes to enable economic prefabrication of the plate blanks by the manufacturer.

The present g1 is equal to the previous d x ½ for round bars. Hence there has been no change except for a slight correction of dimension f1, which affects the handle design. Because of the increasing plate weights and press lengths it was necessary to provide a rectangular bar in addition to the round one, whereby the rectangle should only be regarded as enveloping curve for all possible support profiles with a flat top edge. All members of the committee also considered

DIN 7129 it desirable to specify the dimension g2 in the case of rectangular bars. As further developments can be anticipated in future, however, it was agreed not to specify this dimension in the first instance, but to observe development and, incorporate it in the standard at a later date.

Specification of the position of the slurry feed in the case of recessed plates caused the greatest difficulties. Selection of the dimension $C2$ is dependent not only on the sedimentation tendency of the mixtures to be filtered, and the theoretical views associated with this subject, but also on design factors specific to the manufacturer. For example, the dimensions $C2$ for the plate sizes 1200 and 1300 were disputed to the end. Agreement was reached only with reference to footnote 3 in Table 4 and the inclusion of 2 alternatives, which are necessitated by different production factors.

Physical properties of Hostalen PP grades

The data quoted are average values. Unless otherwise stated, the test specimens were prepared from compression moulded sheet and tested without prior annealing under standard climatic condition 23/50-1 DIN 50014; sheets were manufactured according to DIN 16774 Part2.

No.	Property	Unit	Test method	Test specimen
1.	Density at 23 °C	g/cm³	DIN 53479, method A	sheet, 4 mm
2.	Melt flow index MFI 190/5	g/10min	according to	preform, sintered from
3.	Melt flow index MFI 230/2,16	g/10min	DIN 53735	granules or powder
4.	Melt flow index MFI 230/5	g/10min		
	Mechanical properties			
5.	Yield stress	N/mm²	DIN 53455, testing rate	test specimen 3 with with dimensional
6.	Elongation at yield	%	125 mm/min	proportion 1:4
7.	Flexural stress at 3.5% deflection	N/mm²	DIN 35452, testing rate 5mm/min	80mm x 10mm x 4mm
8.	Modulus of elasticity (3-point bending)	N/mm²	DIN 53457	80mm x 10mm x 4mm
9.	Flexural creep modulus, 1 min value	N/mm²	flexural creep test, $\sigma b = 5$ N/mm²	120mm x 20mm x 6mm upright loading
10.	II 132/30 and II 358/30		test load 132 N or 358 °N	
11.	Shore hardness D, 3 sec value	---	DIN 53505	sheet, 6 mm
12.	Impact strength	mJ/mm²	DIN 53453	small-size standard test bar, injection moulded
	Notched impact strength			
13.	at 23 °C	mJ/mm²	according to test bar,	small-size standard
14.	at 0 °C	mJ/mm²	DIN 53453	with V notch
15.	at - 20 °C	mJ/mm²		
	Izod Impact strength		ISO 180/4A;	
16.	at 23 °C	J/m	ASTM D 256	63.5mm x 12.7mm
17.	at 0 °C	J/m		x 3.2mm,
18.	at - 20 °C	J/m		injection moulded
19.	at - 40 °C	J/m		
20.	Abrasion by the abrader wheel method, Vicat softening temperature	mm³/100 rev	DIN 53754 DIN 53460	sheet

LIQUID SOLID SEPARATION WITH FILTER PRESSES

21.	VST/A/50	°C	in liquid	20mm x 20mm x 4mm
22.	VST/B/50 Heat deflection	°C	DIN 53461	
23.	temperature method A	°C	ob = 1,8 N/mm²	110mm x 10mm x 4mm
24.	temperature method B	°C	ob = 0,45 N/mm²	

Hostalen PP grades

Homopolymers

PPH	PPH	PPK	PPN	PPR	PPT	PPT	PPU	PPU	
1050;	**2250**	1060 F;	1060;	1060;	1070;	1770 F;	1080 F	1080	
1850;	grey 34	1060 F1	1060 F;	1060 F;	1070 S1;	1770 F2		1080 S1;	
2050;			1060 F 1;	1060 F1;	1270			2080 S1	
4050;			1060 F 2;	4160				AST L	
4150			1060 F 3;						
			2060;						
			4060;						No.
			4160						
0.902	0.915	0.903	0.905	0.905	0.907	0.907	0.907	0.907	1.
0.6	0.5	2	4	5	10	13	22	25	2.
0.3	0.5	0.9	2	3	5	8.5	12	15	3.
1.2	1.2	4	9	13	24	33	55	65	4.
31	33	34	35	35	38	-	-	-	5.
16	20	18	12	12	12	-	-	-	6.
26	29	-	30	30	35	40	-	41	7.
-	-	-	-	-	-	-	-	-	8.
1200	1100	1250	1300	1350	1450	1400	1600	1650	9.
63	70	68	70	73	78	85	88	90	10.

70	72	73	70	70	70	72	75	75	**11.**
no failure	no failure	no failure	no failure	no failure	no failure	no failure	15	15	**12.**
10	15	8	7	7	5	4	4	4	**13.**
5	5	-	3	-	2,5	2	-	2	**14.**
4	3	-	2	-	-	-	-	-	**15.**
55	-	35	35	-	25	-	20	-	**16.**
-	-	-	-	-	-	-	-	-	**17.**
-	-	-	-	-	-	-	-	-	**18.**
-	-	-	-	-	-	-	-	-	**19.**
16	15	14	13	14	16	-	15	-	**20.**
150	-	152	152	-	155	150	153	155	**21.**
87	-	88	90	-	98	100	93	100	**22.**
56	57	-	60	62	60	60	-	70	**23.**
110	112	-	110	107	120	110	-	130	**24.**

Hostalen PP grades

					Block copolymers					
PPH 22222 grey 34; 4122	**PPH 1022 1022 F; 1822**	PPN 1034	PPN 1752	PPR 1042; 1042 P; 1042 S1; 4042; 1442; 4242 AST-L	PPR 1745	PPT 1752 S1	PPU 1762 S1	PPU 1734; 1734 S1 AST-L	PPW 1752 s1; 1752 S1	**No.**
0.912	0.900	0.902	0.904	0.903	0.906	0.904	0.905	0.902	0.907	**1.**
0.4	0.6	2.5	2.3	8	6	11	25	25	60	**2.**
0.25	0.3	1.5	1.4	4	3.5	6	14	14	40	**3.**
1	1.3	6	5.5	18	14	28	60	60	150	**4.**
25	27	23	26	28	22	29	30	25	25	**5.**
17	18	13	12	13	10	14	15	11	10	**6.**
22	20	22	29	27	26	28	29	22	30	**7.**
-	940	-	-	-	1200	-	-	-	-	**8.**
900	800	1000	1150	1150	1050	1200	1400	1000	1150	**9.**

50	49	52	55	57	54	57	65	52	58	10.
67	66	-	65	66	65	67	68	-	-	11.
no fail	no fail	no fail	no fail	no fail	no fail	no fail	no fail	no fail	no fail	12.
40	40	no fail	25	20	38	15	10	16	9	13.
15	17	14	11	11	17	8	7	12	16	14.
7	10	9	7	7	11	5	4	8	4	15.
-	830	180	-	120	600	-	-	150	-	16.
-	100	60	-	50	85	-	-	55	-	17.
-	30	40	-	30	55	-	-	35	-	18.
-	25	30	-	-	35	-	-	30	-	19.
15	13	-	-	14	-	-	-	-	-	20.
-	145	143	145	147	137	147	146	144	148	21.
-	73	72	74	75	72	76	78	74	75	22.
53	51	57	55	59	52	60	61	50	55	23.
100	95	105	105	115	100	115	120	100	105	24.

Material Properties NBR Rubber

Polymer	NBR
Stabiliser	non- Staining
Operation temperature	Continuos (1000h) – 20 bis + 120° C
	Short time (10h) + 130° C

Property	Unit	Requirement	Test method
Density	g/cm³	1.20 ± 0.03	DIN 53 479 – A
Hardness	Shore A	70 ± 5	DIN 53 505
Tensile strength	N/mm²	min 15	DIN 53 504
Elongation at break	%	min 400	DIN 53 504
Tear strength	N/mm	min 10	DIN 53 507
Rebound resilience	%	min 30	DIN 53 512
Compression set 22h/70° C/25%	%	max 35	DIN 53 517

SQUEEZE PRESSURE MEMBRANE FILTER PLATES

The following charts are guidelines only and vary from manufacturer to manufacturer

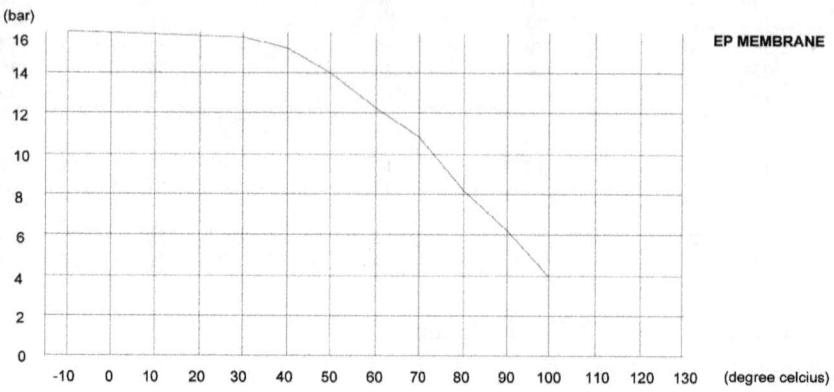

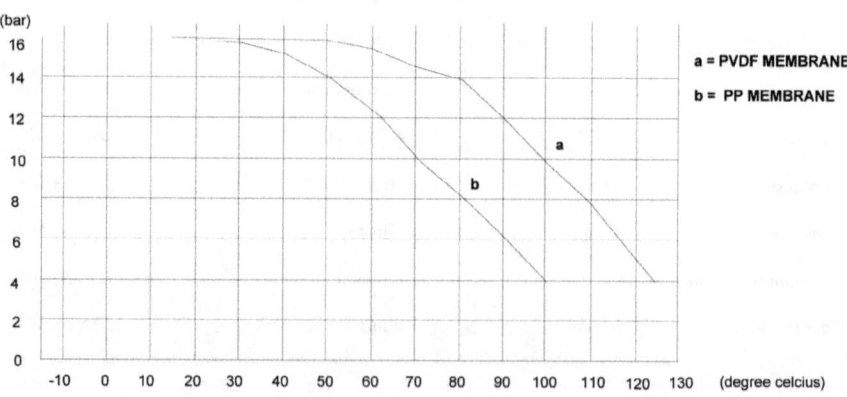

SQUEEZE PRESSURE RISE MEMBRANE FILTER PLATES
(bar/min)

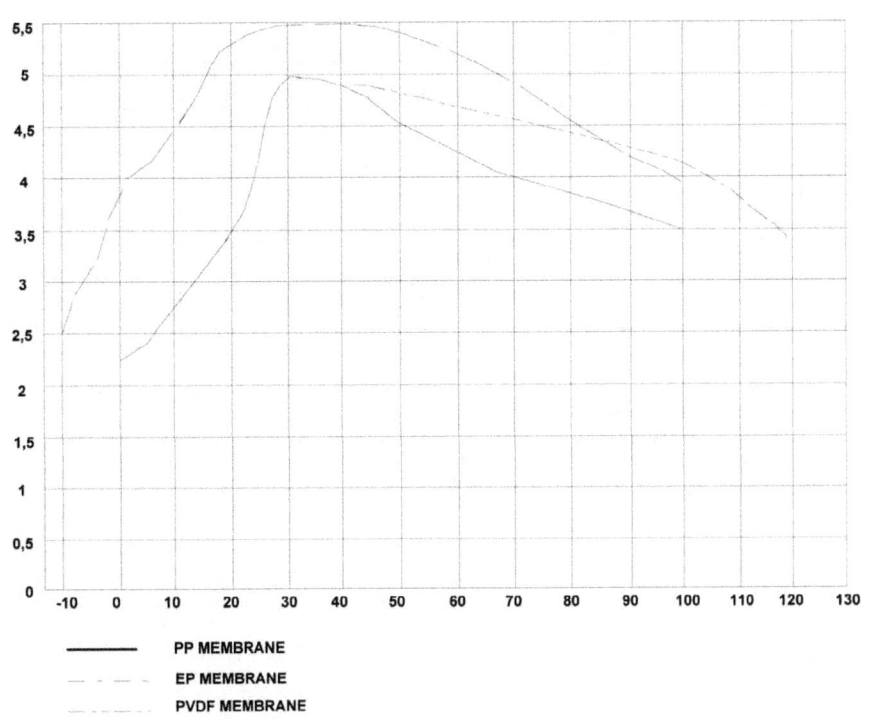

FILTRATION AND WASH PRESSURE MEMBRANE FILTER PLATES

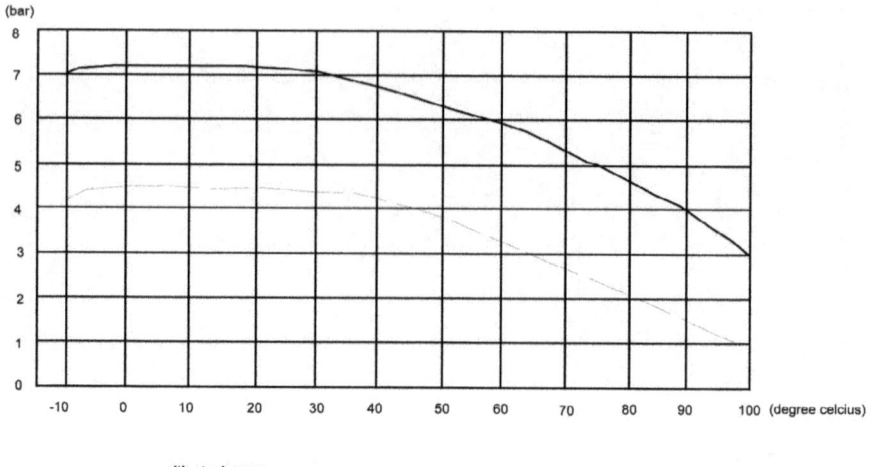

———————— with staybosses
- - - - - - - without staybosses

LIQUID SOLID SEPARATION WITH FILTER PRESSES

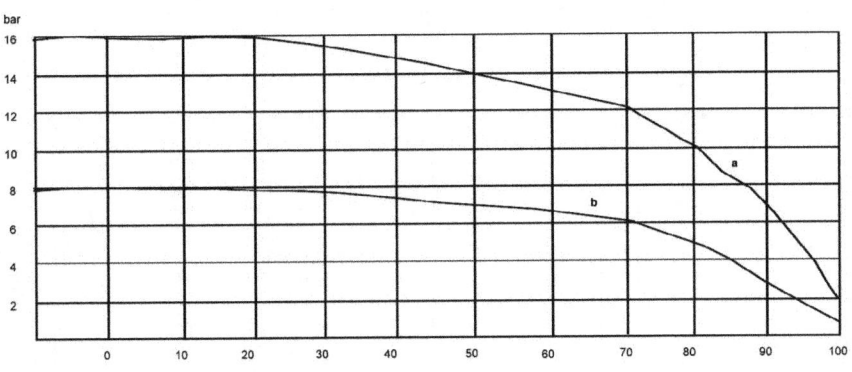

Diagram 1 — Filtration- and washing pressure vs. operation temperature

Line a

chamberplates size 250 - 630 with 4 staybosses (K16)
chamberplates size 800 - 1200 with 4 staybosses web thickness 30 - 32 mm (K16)
chamberplates size 1200 - 2000 with more than 4 staybosses web thickness 37 - 45 mm reinforced web (KA16)
chamberplates size 1500 x 2000 with more than 4 staybosses web thickness 45 mm reinforced web (KA16)

Line b

chamberplates size 800 - 1500 with 4 staybosses web thickness 25-35 mm (K8)
chamberplates size 1200 - 1500 with more than 4 staybosses web thickness 30 - 35 mm reinforced web (KA8)
chamberplates size 1500 x 2000 with more than 4 staybosses web thickness 35 mm reinforced web (KA8)

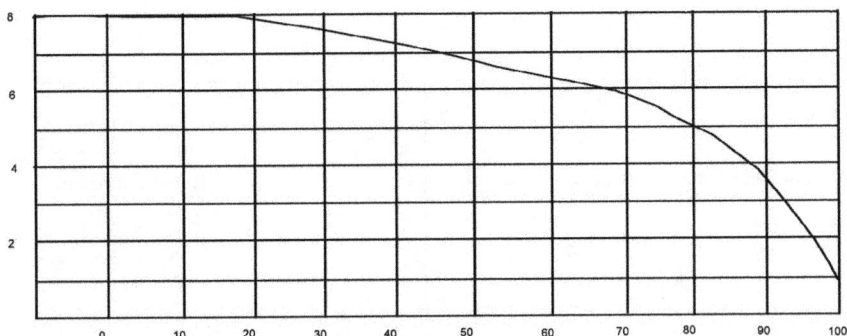

Diagram 2 — Filtration- and washing pressure vs. operation temperature

Standard chamberplates without staybosses (KC and KCE)
Standard filterplates and filterframes

Change of length by warming up / cooling down

Warming up - Elongation

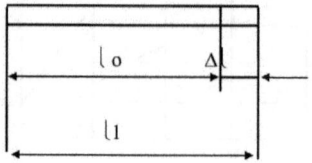

1.) Change in Length Δl

$$\Delta l = l_0 \times \alpha \times \Delta T$$

2.) Length l_1 at warming up

$$l_1 = l_0 + \Delta l$$

3.) Temperature Difference at warming up

$$\Delta T = \frac{\Delta l}{l_0 \times \alpha}$$

4.) Original length l_0

a) $$l_0 = \frac{\Delta l}{\alpha \times \Delta T}$$

b) $$l_0 = l_1 - \Delta l$$

5.) Coefficient of expansion

$$\alpha = \frac{\Delta l}{l_0 \times \Delta T}$$

Cooling down - Shrinking

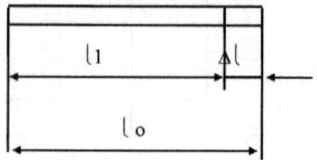

1.) Change in Length Δl

$$\Delta l = l_0 \times \alpha \times \Delta T$$

2.) Length l_1 at cooling down

$$l_1 = l_0 - \Delta l$$

3) Temperature Difference at cooling

$$\Delta T = \frac{\Delta l}{l_0 \times \alpha}$$

4.) Original length l_0

a) $$l_0 = \frac{\Delta l}{\alpha \times \Delta T}$$

b) $$l_0 = l_1 + \Delta l$$

5.) Coefficient of expansion

$$\alpha = \frac{\Delta l}{l_0 \times \Delta T}$$

Legend : l_0 = Length at room temperature (23 °C)
l_1 = Length at warming up or cooling down
ΔT = T1 - T0, at warming up
= T0 - T1, at cooling down
T0 = Basic temperature i.e. 23 °C
T1 = Temperature of the warmed up / cooled down part
α = $1,5 \times 10^{-4}$ K^{-1} = 0,00015 mm/mm/°C for Polypropylene

Calculation of slurry throughput / chamber filling of a membrane filter press Required Data

Type of filter elements, size
Chamber volume (dm^3)
Number of chambers
Total volume filter press (dm^3)
70%—volume filter press (dm^3)
75%—volume filter press (dm^3)
80%—volume filter press (dm^3)
Dry solids content slurry (kg/dm^3)
Specific weight dry solids (kg/dm^3)
Dry solids content filter cake (%)
Specific weight filter cake (kg/dm^3)

Calculation

a) kg dry solids :

$$= \frac{\text{specific weight filter cake}}{100} \times \text{dry solids content filter cake} \times \text{\% press volume}$$

b) kg water

$$= \frac{\text{specific weight filter cake}}{100} \times \text{water content filter cake (\%)} \times \text{\% press volume}$$

c) Slurry throughput (dm³) :

$$= \frac{\text{kg dry solids filter cake}}{\text{kg dry solids slurry}}$$

d) Specific weight dry solids

$$= \frac{\text{kg dry solids filter cake}}{\text{\% volume filter press - water}}$$

e) Proof :

$$\frac{\text{kg dry solids}}{\text{specific weight dry solids}} + \text{kg water} = \text{\% press volume}$$

Example :

Specific weight dry solids : 2,4 kg/dm³
Specific weight filter cake : 1,5 kg/dm³
Humidity of filter cake   : 35%
Specific weight water   : 1,0 kg/dm³
Calculation specific weight filter cake :100

(dry solids cont. : spec. weight dry solids) + (humidity : spec. weight humidity)

$$1,5 = 100 : \{(65 : 2,4) + (35 : 1)\} = 1,61 = \text{wrong !!}$$

Calculated specific weight dry solids in the filter cake :

kg dry solids   : (1,5 : 100) .65 = 0,975 kg
kg humidity (water)   : (1,5 : 100) .35 = 0,525 kg; Volume: 0,525 dm³
specific weight dry solids   : 1 dm³ - 0,525 dm³ = 0,475 dm³;
                                                  : 0,975 kg : 0,475 dm³ = 2,0526 ~ 2,05 kg/dm³

LIQUID SOLID SEPARATION WITH FILTER PRESSES

For the calculation example the following additional figures have been taken into account :

- dry solids cont. filter cake : 65%
- specific weight humidity (water) : 1 kg/dm³
- dry solids content slurry : 0,15 kg/dm³
- chamber volume : 88 dm³
- number of chambers : 96
- total volume filter press : 88 x 96 = 8448 dm³
- 75% - volume filter press : 11760 x 0,8 = 6336 dm³

Example:

Calculation :

a) Kg dry solids :

$$= \frac{\text{specific weight filter cake}}{100} \times \text{dry solids content filter cake} \times \text{\% press volume}$$

$$= \frac{1,5}{100} \times 65 \times 6336 = 6177,6 \text{ kg}$$

b) Kg water :

$$= \frac{\text{specific weight filter cake}}{100} \times \text{water content filter cake (\%)} \times \text{\% press volume}$$

$$= \frac{1,5}{100} \times 35 \times 6336 = 3326,4 \text{ kg}$$

c) Slurry throughput (dm³) $= \dfrac{\text{kg dry solids filter cake}}{\text{kg dry solids slurry}} = \dfrac{6177,6}{0,15} = \underline{41184 \text{ dm}^3}$

d) Specific weight dry solids :

$$\frac{\text{kg dry solids filter cake}}{\text{\% volume filter press - water}} = \frac{6177,6}{6336 - 3326,4} = 2,0526; \quad \text{corresponds with figure example 1}$$

e) Proof :

$$\frac{\text{kg dry solids}}{\text{spec. weight dry solids}} + \text{kg water} = \text{\% press vol.} = \frac{6177,6}{2,0526} + 3326,4 = 6336 \text{ dm}^3$$

The figure corresponds to the 75% press volume.

For the example the calculated recommended slurry throughput per batch is 41,184 m^3.

Attention : The calculated figure (s) have to be proved in practice. Differences in the dry solids content in the slurry have to be as small as possible. Otherwise new calculation is required.

Criteria for Membrane Selection

Membranes represent quite an investment and therefore prior to taking any decision certain criteria should be taken into consideration well balancing the pros and cons of the individual designs that are available.

I. Application Details

The application in which membranes are supposed to be used should be clearly described and in case experience is already available from any past operation of membranes they should be mentioned.

The main topics are:
1. Description of the filtration application.
2. Description of previous experience with membranes and/or with the application.
3. Description of the most important requirements for the intended application.

Based on the above the design, construction, material of membrane and application data can be evaluated and prepared for a first decision.

It goes without saying that any experience with membranes in the same application before is of great help since the fact and figures based on the records during the previous operation of membranes is of utmost importance and shortens the time that is required to take a decision tremendously. Additionally the gained experience can trigger a further in-depth study of possible design and construction changes, if not to say improvements, special material selection or new design/construction in general.

Experience should by no means be underestimated in such a decision finding mission.

II. Monetary Aspects

As mentioned, membranes represent quite an investment. Thus the decision to use membranes for a specific application must consider many aspects which are related with the filtration result, the operation and the work force that is involved as well as maintenance and replacement costs and additional equipment required for the operation of membranes.

The monetary value of a membrane can easily exceed the value of a standard filter element three to six times. There are applications where the filtration result achieved with membranes does not exceed the dry solid content achieved with conventional filter elements and the savings is only to find in the range of filtration time which in some cases might not be of any importance at all. It also has to be considered that in some areas the issue labor costs does not play an important role at all or that the operation skill required does not exist and therefore not recommendable to use membranes at all, unless proper steps are taken to find a remedy for the existing obstacle.

Some decisions are merely based on a non-monetary aspect others strictly on monetary issues.

As such an evaluation can be objective or subjective either on the manufacturers or on the end user's side. Nevertheless a clear understanding should be found between the two parties to avoid any disappointment at a later stage.

A. Monetary Evaluation

The objective part deals with the fact and figures of the process.

Still easy to calculate are the costs for the investment, the installation and the operation.

Rather difficult is the calculation of the savings or the monetary benefits. They are as follows:

1. Waste disposal costs
2. Thermal energy costs in case of incineration or any other form of additional cake drying
3. Operation labor costs
4. Wash media costs
5. Disposal costs for wash media
6. Increase in capacity versus conventional systems
7. Achievement of higher yield
8. Higher dry solid content

Details

Waste Disposal Costs

Whether an existing liquid-solid separation system other than filter press-membrane combination is available or whether an existing conventional arrangement filter press-filter element will be retrofitted the result achieved with the selection of membranes is obvious and can be seen in the higher dry solids achieved and as such in lower costs for the disposal of cake due to less volume and weight to be land filled or incinerated. The detailed costing depends on the individual case and costs for transportation, distance, pre-treatment costs, incineration costs and any additional fees prescribed by the authorities or the communities.

The current costs have to be determined and compared with the savings as well as with the investment. The decision to install membranes surely increases the investment but with increasing investment the savings are increasing in a ratio, which exceeds the investment, costs tremendously.

In case of a conversion from an existing liquid-solid separation system other then filter presses to a filter press—membrane configuration a test filtration is advisable to underline any decision pro such system.

Thermal Energy Costs

Thermal energy costs can apply in wastewater treatment as well as in product filtration where the cake has to be additionally dried before any further processing.

It actually does not require any explanation, less energy costs for the evaporation of liquid equals less BTU for the incineration.

Here the calculation of the monetary advantages is rather simple and the production manager in charge should have the figures off hand.

Operation Labor Costs

In case of a retrofitting of an existing filter press—conventional filter element configuration into filter press – membrane system, i.e. replacing plate and frames or chamber plates the savings are to be found in many areas:

Less time for the handling of the same quantity of sludge with equal or even higher result in the dry solids. The same volume per day will require less hours of operation.

The decision is to expect that the volume per day will be increased. The savings are obvious.

The membrane system is to increase the dry solid content. Costs for cake removal are decreased, the removal time is faster, no additional time required to scrap off the cake from the filter cloth and also less volume to handle downstream of the filter press.

Even more effective is the cake removal compared with plate and frames.

Not to mention the rather messy affair of cake removal on a plate and frame configuration, the time to scrap off remaining cake from the cloth and the danger of damaging filter cloth is entirely eliminated.

Wash Media Costs

The savings of wash media only refers to those applications where the cake has to be washed prior to any disposal. The main aim is to replace the mother liquid in the cake with water or any other wash media. This can be the removal of soluble salts or other liquids, which is done by rinsing some sort of leaching. These applications are found in the dyestuff, mining and chemical industries.

Regardless which industry it is it is obvious that the savings in wash media costs are tremendous underlined by the following arguments:

Application of a low-pressure squeeze before and/or during the wash step avoids that the cake becomes receptive to channeling during the wash process.

Application of a low-pressure squeeze before and/or during the wash step leaves less mother liquid in the cake that has to be washed.

Application of the final squeeze results in recovery of additional wash media.

Disposal Costs for Wash Media

Both cases are counting, the disposal of wash media into the municipal sewer based on costs per cubic meter or the required additional and separate treatment of the wash media by means of a separate treatment plant due to the nature of the wash media and the non-permission to disposal at such stage.

Increase in Capacity

The demand for increased capacity usually results in an investment in an additional filter press.

By using membranes this investment can be abolished and savings are easily calculated. In case of a full membrane application combined with a new filter press the membranes represent approximately 50 % of the total investment.

Further increasing capacity results are due to reduced filtration time.

The result as mentioned earlier is:

Less time for the same volume or same time with more volume

Achievement of Higher Yield

In the application of palm oil the usually known dry solid content in liquid-solid separation is replaced by the terminology YIELD. In the separation of olein and stearin the results normally range from 76-80 % yield. Each percent of yield depending on the country represents an amount that cannot be neglected.

With the right selection of membranes and the right selection of design and construction as well as material a valuable contribution to a maximum yield can be given.

In terms of yield special production goals towards quality of olein have to be taken into consideration whereby the high yield might not be so decisive but the afore mentioned does apply in the majority of the palm oil applications.

Higher Dry Solid Content

The advantages by producing a higher dry solid content are actually the main issue of the membrane application combined with the fact that in such cases where membranes are used the filtration time or cycle time is tremendously reduced at the same time. In the majority of cases both higher dry solid content and reduced cycle time goes hand in hand.

Nevertheless many aspects have to be taken into consideration, the main issues were listed above but additionally floor space, feed pumps, plumbing, conveyors, labor etc are further topics that need to be watched.

Lastly the combination of a mixed pack, i.e. chamber plate/membrane plate combination should be investigated and might be in some cases the better choice with savings of approximately 25 % in comparison with a pure full membrane pack.

Special attention however is here required when the type of membrane is selected.

B. Non-monetary Evaluation

The non-monetary evaluation is not objective, calculations of advantages versus disadvantages; savings etc. are not possible to the extent as they are possible for the monetary evaluation.

The following are to be considered as subjective:

1. Higher quality in the end product

2. Possibility to meet increasing environmental control
3. Flexibility within the process (different products)

Higher quality in the end product can lead to advantages over the competitors and can also lead to increasing sales price. For example the volume of wash media to remove soluble salts in the product during filtration might be above permitted level. By designing the right membrane system the limited volume could be accommodated which would result in a better quality of product.

Possibility to meet increasing environmental control refers to the fact that existing laws from the individual governments re disposal of hazardous waste, regulations for landfill etc are still based on other systems than filter press—membrane configurations. Under the existing rules and regulations are many other systems to find which judged by the achievement of dry solids are far exceeded by the application of membranes. The investment in membranes would therefore not only reduce the possibility of an installation of a new system in case the rules and regulations are changed again it would also give ample safety in terms of forthcoming changes of rules and regulations since the indicated standards by the law are more than met.

Flexibility within the process is important in case more than one product or within one product various parameters have to be covered which a well-designed membrane, which meets all the requirements, can easily do.

III. Construction Evaluation

IV. Application Evaluation

III and IV are summarized in the following breakdown. For construction evaluation the explanations are categorized with III (n) and for application evaluation with IV (n).

In a further table the results out of the evaluation are summarized and form the basis for the later decision on the type of membrane to be selected.

The indicated figures III (n) and IV (n) can refer to more than one category.

III-1 III-2

List chemicals with concentration in the sludge, the solids and the temperature.

The use of the chemical resistance list is necessary; especially then when no previous experience is available.

Basic guidelines to be considered are:

Any traces of manganese or copper and cobalt require copper stabilized Polypropylene.

Chlorine oxidizes Polypropylene and turns it brittle.

EP contains rubber and Polypropylene. Both chemical resistance characteristics apply.

For rubber membranes the body plates can be of different material, like Polypropylene, Cast Iron, Aluminum therefore the influence of the sludge on the base plate material has to be investigated.

Temperature is decisive for the design of the base plate or rather web thickness including the thickness of the membrane.

III-1 III-6

Is the filter press to be used for more than one application and if so, check on the chemical resistance for the additional products.

As the individual product for the application is examined in terms of chemical resistance of the membrane material,

the filtration process etc. the same has to be done with any additional product. After this step is done it has to be elaborated whether all the criteria are met with the chosen membrane or whether it is not advisable to use a separate filter press—membrane installation for the additional products.

III-1 III-2

Check on the chemicals that are used as wash media and the temperature.

The chemical resistance based on the wash media is not prevailing it may be very important to know the difference in temperature for the process as well as for the washing. According to these findings the material for the membrane has to be selected.

III-2 III-4 III-10 IV-13

Is previous experience available and what are the pressures for feeding, squeezing etc.

Feed pressure and temperature is important to know to determine the design and the construction of the membrane as well as the body plate of the membrane. The main aim of the membrane is to utilize low feed pressure resulting in high flow rate, use higher pressure for squeezing and then to continue with the next filtration.

There are limited cases where the feeding is done at 10 bar followed by a squeezing of 14 bar. These limited cases should be considered when the base plate of the membrane is designed and when the material for the base plate is selected.

In case of retrofitting the maximum allowable pressure of the filter press has to be considered. There are cases where even with low pressure filter presses the application of membranes at low-pressure operations are recommendable and the investment in a new filter press can be avoided.

Out of balance pressures have to be considered in view of the design and construction of the base plate/web thickness.

PVDF is a material, which does not only apply for higher temperatures and shows a superior chemical resistance but also offers a higher physical strength, which is to be considered when high pressures are involved. The basic high material costs of PVDF however have to be evaluated.

III-2 III-4

Does the temperature around the filter press change.

This section might be exaggerated but there are cases where the temperature surrounding the filter press does change and consequently has an influence, not on the application but on the material of the membrane.

Mobile filter presses for example being exposed to outside temperatures do have to use a selection a membrane material which can resist low temperatures for example in winter time. In that case EP material or rubber are the correct material to chose over Polypropylene, which might become brittle. But possible oils in sludge have to be examined.

III-3

Is previous experience available with the various materials for membranes for this specific application?

Apart from the possibly existing experience always check on the piping at customers' side. If in the piping system Teflon gaskets are used the use of gaskets in the membrane has to be re-considered. If the customer is using stainless steel pipes the decision whether to use a base plate of mild steel has to be revised since there might a reason for using stainless steel pipes.

III-8 III-9

Is previous experience available with replaceable or one-piece membranes?

If the performance of one-piece membranes has been without any failure a change to replaceable membranes is not advisable. If failures of the one-piece membrane were in the range of above 10% and were not related to chemical attack the recommendation to use replaceable membrane is a logical step. It is also recommendable to use replaceable membranes in case of a high amount of cycles per day and high wear and tear on the membrane material.

III-4

Which cycle time and how many cycles per day are expected.

Where filtrations are not frequently done a replaceable membrane is totally out of question, however, important where many filtrations at short cycle times are the case.

In applications which due to the temperature would not allow the selection of a certain material for the membrane it might be acceptable to still use this material provided the cycle time is short and the filtration not frequently done.

III-4

Is the filter press a fixed installation or a mobile one?

The best solution is a one-piece membrane but is has to be taken into consideration that due to surrounding temperatures other material than Polypropylene is recommendable. In case of many applications i.e. different products the use of a replaceable membrane might be taken into consideration but it also has to be observed that the maintenance of a replaceable membrane is not exactly what the end user wishes, i.e. replacement of gaskets, cleaning of the membranes due to product getting in between the membrane and the base plate and corrosion of metallic components like center feed ring.

III-5

Is previous experience available with any leakage problems?

If the membrane in case of a decision for a replaceable one is not properly maintained leakages might occur.

These leakages can occur when solids get between the membrane and the base plate. There is also the possibility that squeeze liquid might enter into the filtrate holes.

III-6

Which metallic materials are acceptable under consideration of the chemical resistance list?

Again proper examination of possible chemical attacks on the material has to be observed. In some cases it might be advisable to use hasteloy locking rings.

III-8

Is previous experience available with the selection of materials?

III-8 III-9

Is previous experience available with design and construction of the base plate/ web thickness?

III-10 IV-8

In case of retrofit, how many plates can be inserted, what is the maximum pressure of the filter press and which total weight of filter plate pack is permissible.

LIQUID SOLID SEPARATION WITH FILTER PRESSES

The technical specification of the filter press to be retrofitted has to be carefully examined. In some cases the use of membranes might be totally out of the question due to the maximum pressure of the filter press in other cases the possibility to use membranes might be within the permissible range and can contribute to tremendous improvement in the filtration result even eliminating the need for a replacement of the filter press.

IV-1

What is the approximate solids concentration of the slurry?

This data is already important for the selection of the infeed system whether corner feed or center feed.

More than that, specific gravity is important to be considered when it deviates from that of water.

The diameter of the corner feed is smaller than the diameter of the center feed. Thus slurries with high solids are more likely to block the corner feed than the center feed. Furthermore products that have a tendency to crystallize can easily contribute to uneven cake built-up in the chambers since the crystallization can block the flow of the sludge into the chambers since the infeed slot between two chambers and considering the thickness of filter cloth and assuming no filter cloth locking device is used is too small. Some chambers might be overfilled some under filled a situation, which will cause plate bending, or even breakage of the base plate of the membrane.

In general for slurries above 5 % by weight center feed should be used.

IV-2

Description of the physical nature of the solids, i.e. granular, crystalline, compressible, firm, slimy, porous etc

In case the sludge contains fibrous solids center feed would be the right choice.

Compressible and slimy products are related to a thin cake and have a tendency to clog the filter cloth. A mixed pack should be avoided except when the combination consists of membrane and filter plate (flush plate).

Also higher feeding pressure is required and/or longer filtration time. The possibility of "out-of balance pressure" is given and the most preferred solution would be center feed and design of membranes with stay bosses as support.

Also for center feed are to be considered those slurries with a high viscosity. Another argument for center feed is the possible "out-of balance" situation for solids with a high settling rate.

IV-3

What is the flow rate of the slurry?

Decisive for the selection of corner feed or center feed is the cross sectional area of the feed opening as related to the speed due to the flow rate. The flow rate through the feed channel, which is in the beginning at maximum speed, has to be checked. It is recommendable to use center feed when the flow rate is high in order to avoid possible pressure shocks, which can be easily caused in case of corner feed and at high speed.

Low flow rates and flow rates in the medium range are the basis for either center feed or corner feed.

IV-4

What is the cake thickness before squeeze and after squeeze?

To know the cake thickness before and after squeezing is important for the selection of membranes or mixed pack. Membranes can handle any kind of cake thickness i.e. with the combination of other filter elements i.e. filter plates or chamber plates.

A possible combination is a standard membrane cake thickness with a chamber plate having a cake thickness lower than the membrane.

IV-5 IV-6

Do the solids require a washing. What is the aim of washing?

There might be a number of arguments for one or the other system, like corner feed for washing, no stay bosses etc but these arguments mainly go back to previous experience or developed philosophy. Basically there is no problem at all to use center feed for washing. The advantage of membranes for washing is to be found in the fact that due to pre-squeezing possibilities a channeling of the wash media is avoided. Pre-squeezing also avoids the excessive consumption of wash media and lastly the final squeeze increases the amount of wash media recovery. The decision whether corner feed or center feed is difficult without considering all other aspects that lead to the decision on the type and design of membrane to be selected.

IV-6

Is there any experience available pertaining to washing?

IV-7 IV-11 IV-12

Is there any experience in connection with the selection of the infeed whether corner feed or center feed?

The same question was raised before but as per the following table the answer falls under a different category, which forms only one part of a list of categories that are to be taken into account. Customers' information on experience, the economic, interchangeability within the filter press installations and experience in similar application are all accounting for an easier and confident selection of the right membrane.

In another example the frequent exchange of a barrel neck cloth speaks pro center feed since the infeed diameter is bigger and therefore the installation of the cloth much faster.

IV-9

Which air supply is available for the squeezing system?

For the selection of squeeze media the basic decision is that pressure above 7 bar applies water as squeeze media and below air as squeeze media. But also the costs have to be considered. Air is compressible and as such safety has to be looked at closely. Water is for sure the squeeze media, which provides safety. Any design of the squeeze media supply has to follow the operation manual of the membrane filter elements to eliminate any pressure shocks. An internal squeeze system is no doubt a clean solution however causes problems when any leakage has to be detected. Special attention has to be paid to the installation of the filter cloth since the openings for the squeeze system on the filter elements cannot be blocked. Any squeeze system with water should be accompanied by a vacuum pump to evacuate the membrane after completion of the squeezing step. This evacuation prepares the membrane for the next filtration and minimizes also pressure shocks that could possible occur from the feed pump. Water also has the advantage over air in terms of costs and can be re-used. The flow can be controlled correctly since water is not compressible. The same refers to pressure and release of water into the squeeze system.

In spite of all expectations it is theoretically possible that the hydraulic fails to function and releases the closing force during he squeezing step in such a situation again water is the safest way.

The possibility to squeeze with product, i.e. squeezing with olein in palm oil application and its advantages should not be excluded. Also important in case of squeezing with water the temperature is important for the selection of the material for the base plate, i.e. homopolymer or copolymer. Temperature of water for squeezing below 0 degree celcius has to be taken into consideration.

IV-11 IV-12

Is there any condition of the application that might affect the design of the membranes?

Membrane Web Design Criteria	PP PP One-piece	PP PP Detachable	EP PP One-piece	EP PP Detachable	Comments
III-1 Chemical					
III-2 Physical					
III-3 Experience					
III-4 Service Conditions					
III-5 Leakage					No contamination
III-6 Metals					Chemical resistance
III-7 Economics					Cost effective
III-8 Special Notes					
III-9 Special Notes					
III-10 Filter Press					Mobile, fixed
Overall					

Membrane Web Design Criteria	Rubber PP Detachable	Rubber AL Detachable	Rubber ST Detachable	PVDF PVDF One-piece	Comments
III-1 Chemical					
III-2 Physical					
III-3 Experience					
III-4 Service Conditions					
III-5 Leakage					No contamination
III-6 Metals					Chemical resistance
III-7 Economics					Cost effective
III-8 Special Notes					
III-9 Special Notes					
III-10 Filter Press					Mobile, fixed
Overall					

LIQUID SOLID SEPARATION WITH FILTER PRESSES

	Corner Feed		Center Feed		Air Squeezing		Water Squeezing		Product Squeezing		Comments
Criteria Data	Mixed Pack	Full Pack	Mixed Pack	Full Pack	Int.	Ext	Int.	Ext.	Int.	Ext.	
IV-1 Solids %											
IV-2 Solids nature											
IV-3 Flow rate											
IV-4 Cake Thickness											
IV-5 Washing											
IV-6 Washing Experience											
IV-7 Application Experience											
IV-8 Squeeze Pressure											
IV-9 Air Supply Pressure											
IV-10 Economics											
IV-11 Special Conditions											
IV-12 Special Conditions											
Overall											

ABBREVIATION KEY	
NA	NOT APPLICABLE
NR	NOT REQUESTED
R	REQUESTED
A	APPLICABLE
P	PREFERRED

EXAMPLE FILTER PRESSES FOR TREATMENT OF SLUDGES

GENERAL

For the dewatering of effluents and sludge from industrial and municipal waste treatment the application of filter presses among other filtration/separation devices is a common practice.

Previous applications of filter presses especially during the time plate and frame or only chamber plates were available were not favored, however due to the fact that filter presses can operate with automatic features they are often considered although it is worthwhile to pay attention to other available systems.

Common applications are:

- Blow down sludge from cooling towers
- Sludge from metal
- Alum sludge
- Secondary biological sludge
- Sludge containing oil
- Brine sludge

The solution to dispose the dewatered sludge in landfills is not preferred anymore and due to high moisture content in the dewatering sludge at the time when belt presses, screw presses or rotary drum filters were used the idea of incineration became very costly due to the high demand of energy required for the incineration. In addition to that the transportation cost for the dewatered sludge due to the weight as a result of high moisture content were facts, which were not in favor of the results achieved with the equipment indicated. Among all the filtration/separation devices only the decanter seems to be on a competitive level in view of the dry solids obtained while for both filter press and decanter the operation and maintenance costs need to be observed carefully. Differences can be significant.

FILTER PRESSES

Originally the filter press available some 150 years ago is simple steel construction with a head and end piece, two sidebars or overhead beams. Side bars or overhead beams are connected with the head and end piece. Using the sidebars as support or the overhead beams as basis to fix the filter elements, the filter elements are installed between the head and end piece. The selection of filter elements can be plates and frames (in general not to recommend) chamber plates or membrane plates as alternative a so-called mixed pack, an arrangement of chamber plates and membrane plates as alternate plates.

The results that can be achieved will not only depend on the sludge that has to be treated but also on the decision on which type of filter element will be used.

Deviations in dry solids of up to 20% are possible.

The plate itself consists of a PP plate with a bibbed or grooved surface, the filtration area and inlet and outlet that can be arranged according to the requirements. In general center infeed with four outlets is required, however corner feed is an alternative.

On top of the filter plate a filter cloth is installed which actually is mounted over the two-filtration surfaces. At the feed mostly a so- called barrel-neck is used, as an alternative a filter cloth-locking device can be installed to fix the filter cloth in the feed area. Barrel neck for cost reasons is to recommend. The filter cloth will be fixed on the side of the filter elements in order to have a straight positioning.

A pump is used to feed the sludge into the filter press, i.e. into the chambers between the filter elements through the infeed along the total length of the filter elements arrangement until the chambers are full. Since pressure is applied the solids starts to be collected between the chambers and form a layer, which becomes thicker and forms the so-called cake. After the required cake thickness, which is determined by the decision on the filter element to be used, is formed the dewatering procedure enters into the final cake consolidation to reach the maximum dryness in cake.

During the last mentioned process mores solids are pumped into the filter element arrangement and by doing so the liquid in the cake is displaced. The filtration is completed when liquid stops to flow out. Now the feed pump is inactivated and the pressure in the filter element package is released.

The cake is discharged. Before the cake is discharged the hydraulic closure system is activated to open the end piece of the filter press and depending on the design of the filter press the filter elements are shifted one by one or in groups of five giving enough space for the cake to discharge.

Plate shifters can operate hydraulically, pneumatically or electrically. In terms of time the plate shifting of 5 plates in a group can be done in approx. 1 minute and less.

After the cake is discharged the end piece of the filter press is pushed to its locked position and the next filtration can begin.

In some cases it might be necessary to clean the filter cloth which can be done by either built-in cloth washing devices or by use of external high pressure pumps at up to 100 bar.

LIQUID SOLID SEPARATION WITH FILTER PRESSES

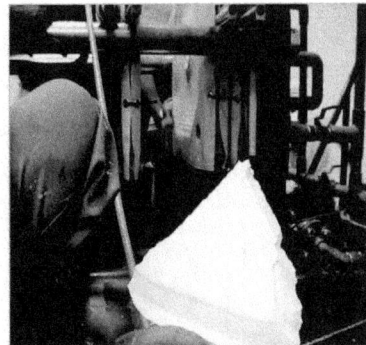

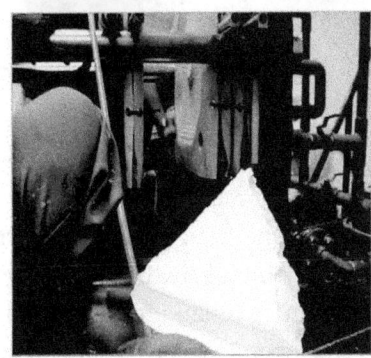

A typical set up of the necessary equipment for sludge dewatering is:
- Sludge collection vessel
- Pump
- Storage vessel for the chemicals that might be required
- Metering Pump
- Conditioner
- Pump for feeding the filter press
- Filter Press
- Arrangement for the cake discharge
- Arrangement for the filtrate discharge

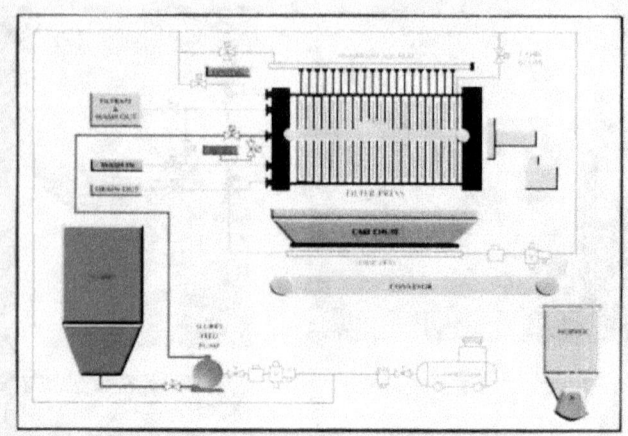

As mentioned earlier the sludge can be different and thus the filtration time. As a basic rule the higher the concentration of the sludge feeding the filter press the shorter the filtration time.

For metal hydroxide an average of 1-3 hours or alum 4-6 hours and for municipal sludge in general 1.5—2.5 hours.

Depending on the selection of filter elements the dry solids can go up to 55%-60%. As a guideline the following dry solid content:

Brine Sludge	60% - 70%
Lime Sludge	30% - 60%
Oil Refinery Sludge	40% - 50%
Alum	25% - 30%
Biological Waste Activated Sludge	30% - 40%
Metal Hydroxide	35% - 55%

SLUDGE CONDITIONING

Apart from the correction of the pH value many industrial sludges can be treated without any prior treatment. Biological waste has to treat with chemicals or requires conditioning of the sludge to prepare the sludge for a filtration process.

The physical process describes heat treatment and freezing of the sludge.

Another step is the application of bulking agents, i.e. filter aid, which can be D. E. (diatomaceous earth), fly ash or similar products. The chemical process is more economical. It is common to use the following conditioners for sludge treatment:

Ferric Chloride

Available in liquid form (39% solution) stored in corrosion-resistant vessels, water is added and mixed to achieve a solution of 10%—15%.

Alum

Available also in liquid form with a solution of 48% mixed with water to reach a solution of approx. 10%

Lime

Available in bags. Filled in a vessel where so-called milk of lime mixture of approx. 10% is achieved by mixing with water.

Polyelectrolyte

Also called polymer can be obtained in form of powder or liquid. Polymer has to be mixed to achieve a solution of 0.1%—0.2%.

Which conditioning agents are to be used?

Biological waste activated	Ferric chloride, Alum or Polymer
Oil refinery	Ferric chloride or Lime
Water Treatment Alum	Lime
Minerals	Polymer

The vessel size for the conditioner should correspond to 20 minutes pumping capacity. The vessel mixer should have a low speed less than 500 ft/min.

Selection of pumps is another criteria.

The main task of the pump is to feed a high volume at a low pressure and reduce to low volume maintaining a high pressure during the later stage of the filtration process.

Suitable are piston pumps (hydraulically driven), double diaphragm pumps (air driven) piston membrane pumps or progressive cavity pump with speed control. The first three pumps have to include a surge suppression vessel installed on the discharge side of the pump to avoid pressure surges on the filter.

For oily and very gelatinous sludges a pre-coating with D. E. is recommended to avoid blinding of the cloth and also to provide a better cake release.

Required are a pre-coat vessel and approx. 10—15 minutes are required to pre-coat the material on the filter cloth before the feeding of the sludge can start.

Comparison decanter, belt press and filter press equipped with membranes based on waste water/dry solid content and polymer flocculent consumption:

Dry solid content (approx.)

Belt press	21% - 23%
Centrifuge	30% - 35%
Membranes	30% - 35%

Polymer flocculent (approx.)

Belt press	3 – 16 g/kg cake
Centrifuge	8 – 10 g/kg cake
Membranes	4 – 12 g/kg cake

LIQUID SOLID SEPARATION WITH FILTER PRESSES

VALVE CHARTS - CORNER FEED PRESSURE

SIMPLE FILTRATION - FIXED VOLUME

	A	B	C	D	E	F	G	H	J	K
FILL/PACK	0	-	-	-	-	X	X	0	-	-
PRE - SQUEEZE	0	-	-	-	-	0	0	0	-	-
WASH										
SQUEEZE										
AIR BLOW										
CORE BLOW										
VENT										
LINE BLOW										

SIMPLE FILTRATION - MEMBRANE

	A	B	C	D	E	F	G	H	J	K
FILL/PACK	0	-	-	X	X	X	X	0	X	0
PRE - SQUEEZE	0	-	-	X	X	0	0	0	X	0
WASH										
SQUEEZE	X	-	-	X	X	0	0	0	0	X
AIR BLOW										
CORE BLOW	X	-	-	0	0	X	X	X	0	X
VENT	X	-	-	X	X	0	0	0	X	0
LINE BLOW	X	-	-	0	X	0	0	X	X	0

WASHING/AIR BLOWING - FIXED VOLUME

	A	B	C	D	E	F	G	H	J	K
FILL/PACK	0	X	X	-	-	0	0	0	-	-
PRE - SQUEEZE										
WASH	X	0	X	-	-	X	X	0	-	-
SQUEEZE										
AIR BLOW	X	X	0	-	-	X	0	X	-	-
CORE BLOW										
VENT										
LINE BLOW	X	X	0	-	-	0	0	X	-	-

WASHING/AIR BLOWING - MEMBRANE

	A	B	C	D	E	F	G	H	J	K
FILL/PACK	0	X	X	X	X	0	0	0	X	0
PRE - SQUEEZE	X	0	X	X	X	X	X	0	0	X
WASH	X	X	X	X	X	X	X	0	0	X
SQUEEZE	X	X	0	X	X	X	0	X	0	X
AIR BLOW	X	X	0	X	0	X	X	X	0	X
CORE BLOW	X	X	X	X	0	0	0	X	X	0
VENT	X	X	0	X	X	0	0	X	X	0
LINE BLOW	X	X	0	X	X	0	0	X	X	0

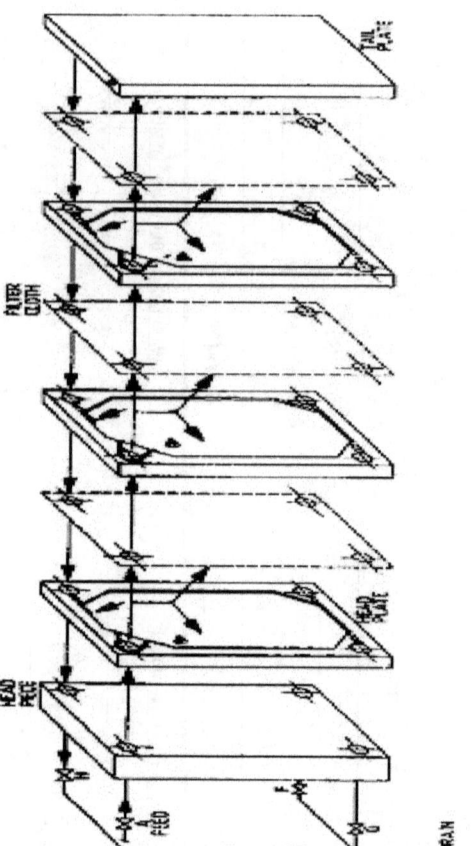

LIQUID SOLID SEPARATION WITH FILTER PRESSES

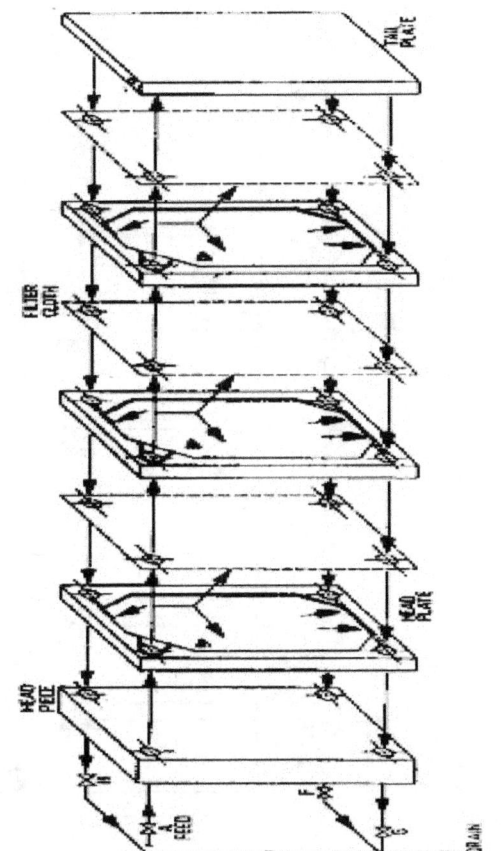

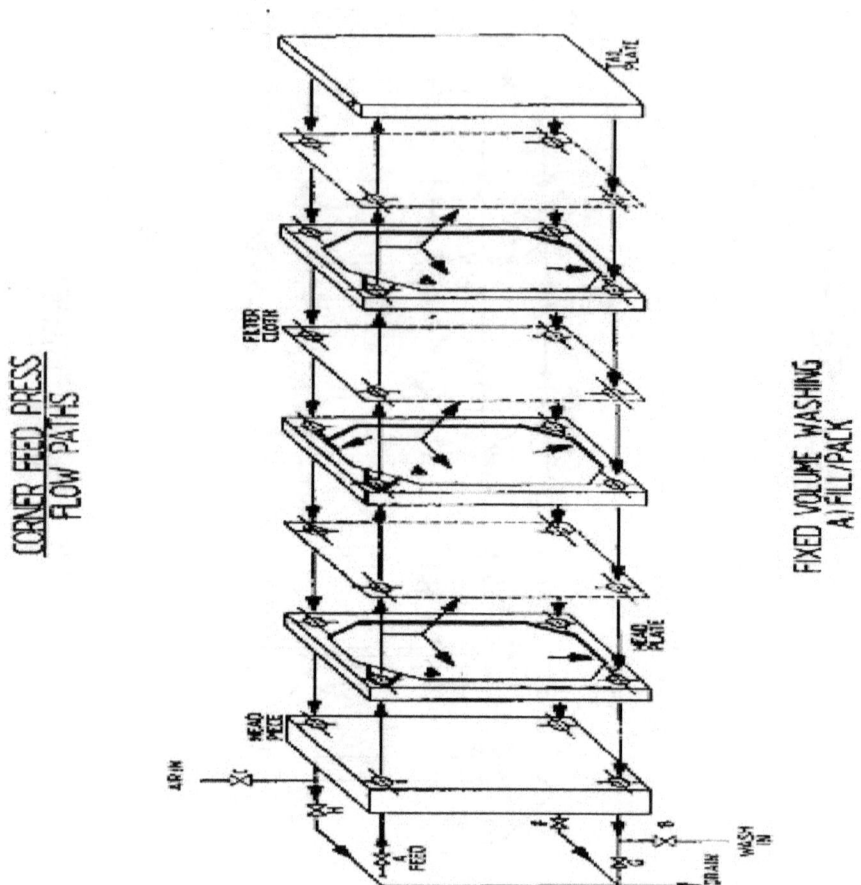

LIQUID SOLID SEPARATION WITH FILTER PRESSES

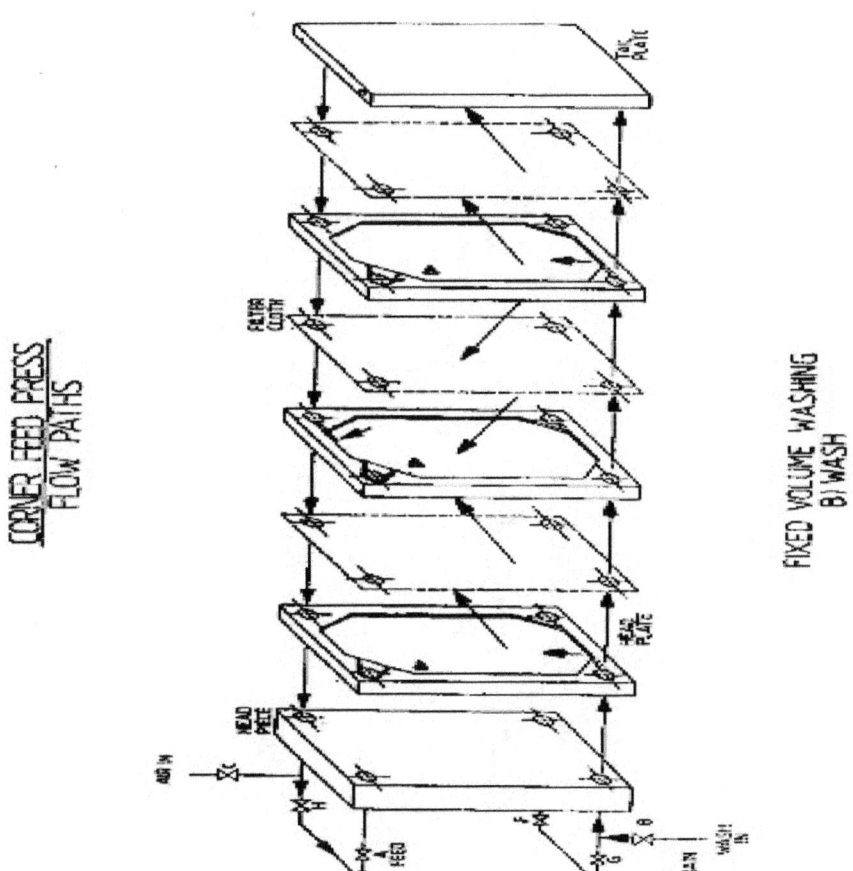

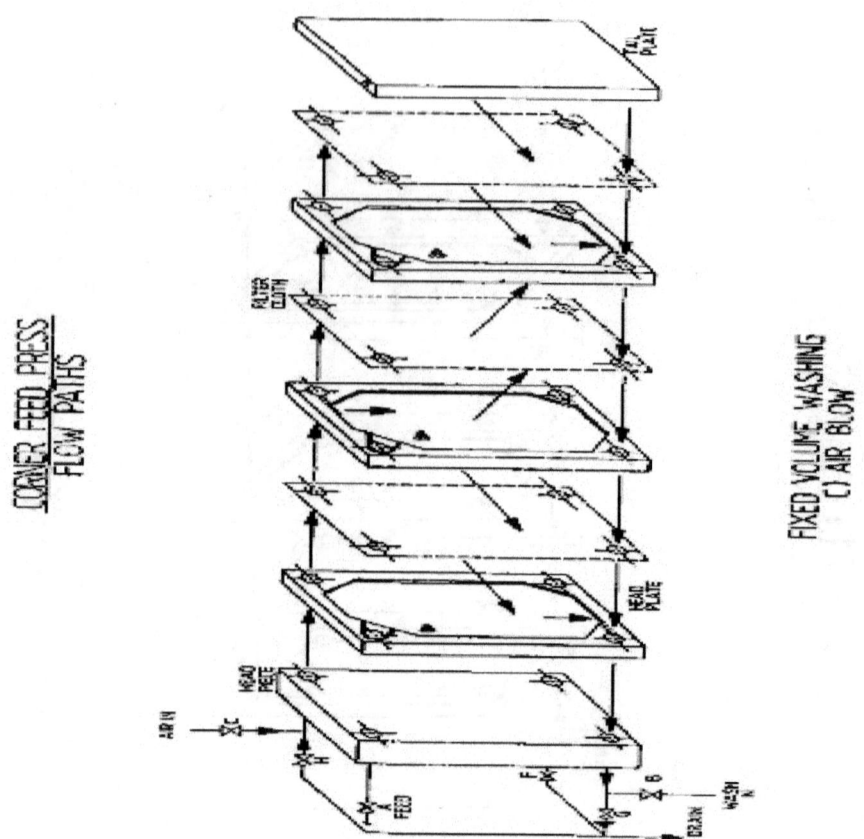

LIQUID SOLID SEPARATION WITH FILTER PRESSES

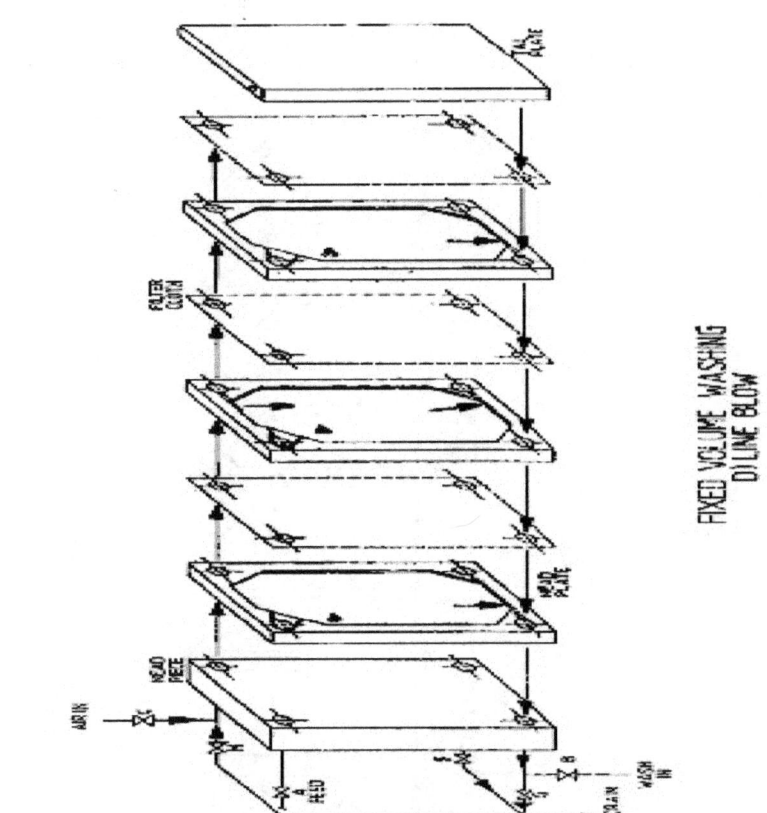

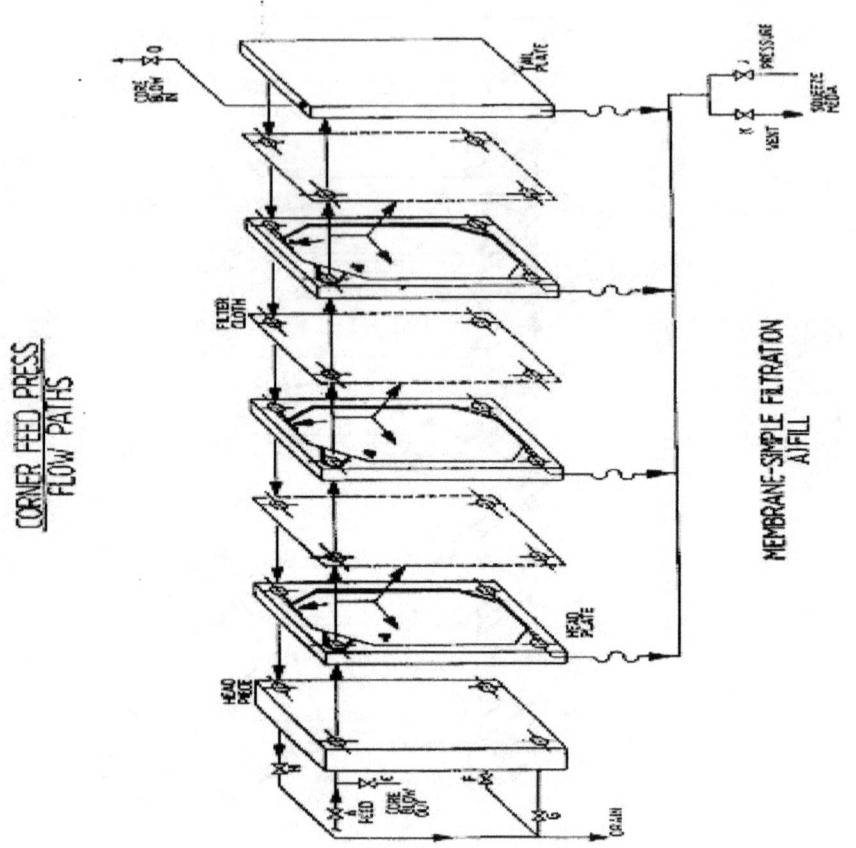

LIQUID SOLID SEPARATION WITH FILTER PRESSES

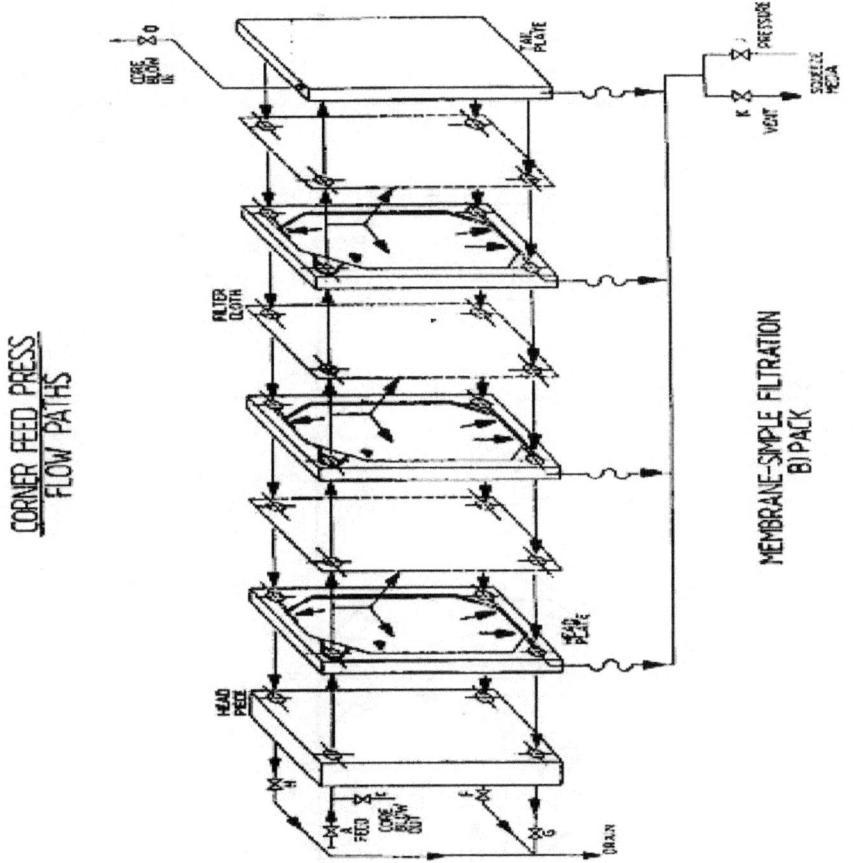

HANS G. VALERIUS PHD

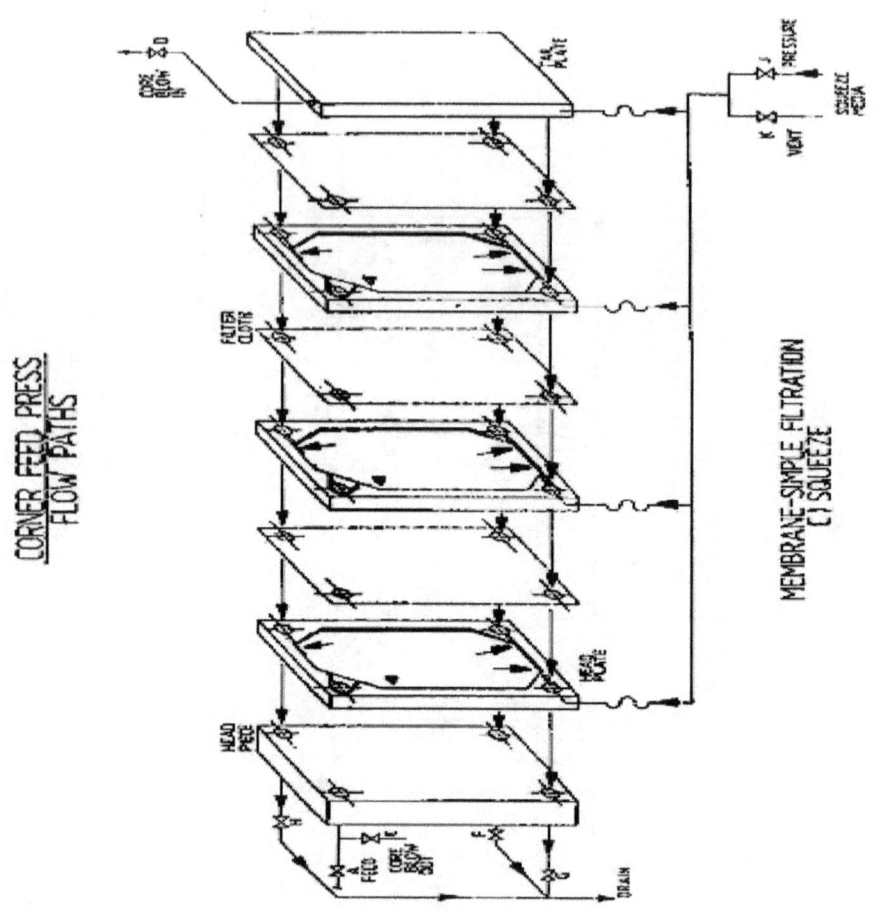

LIQUID SOLID SEPARATION WITH FILTER PRESSES

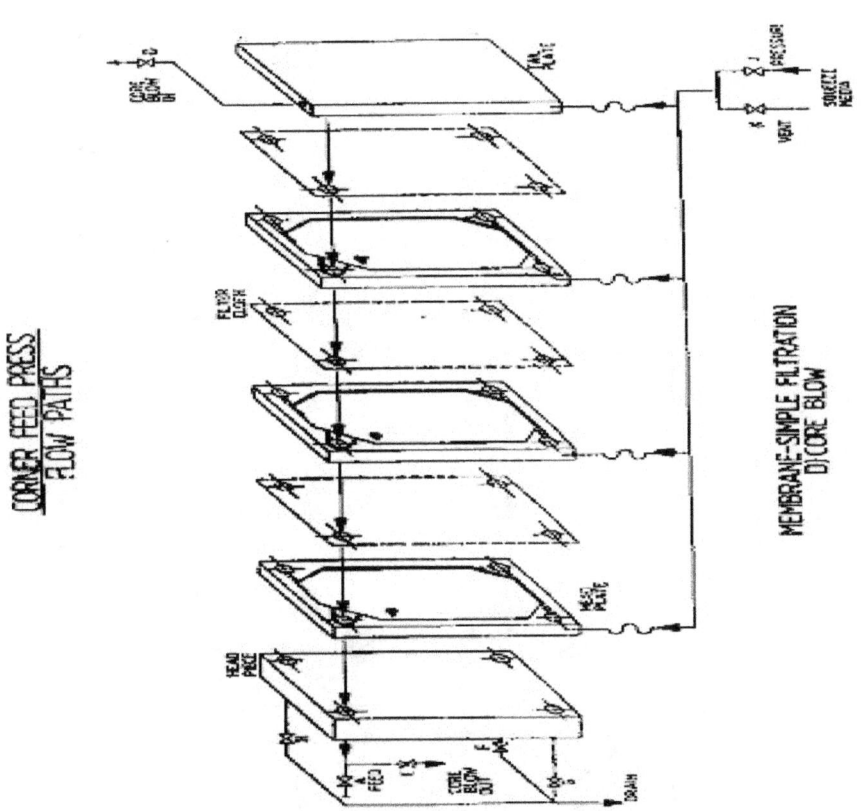

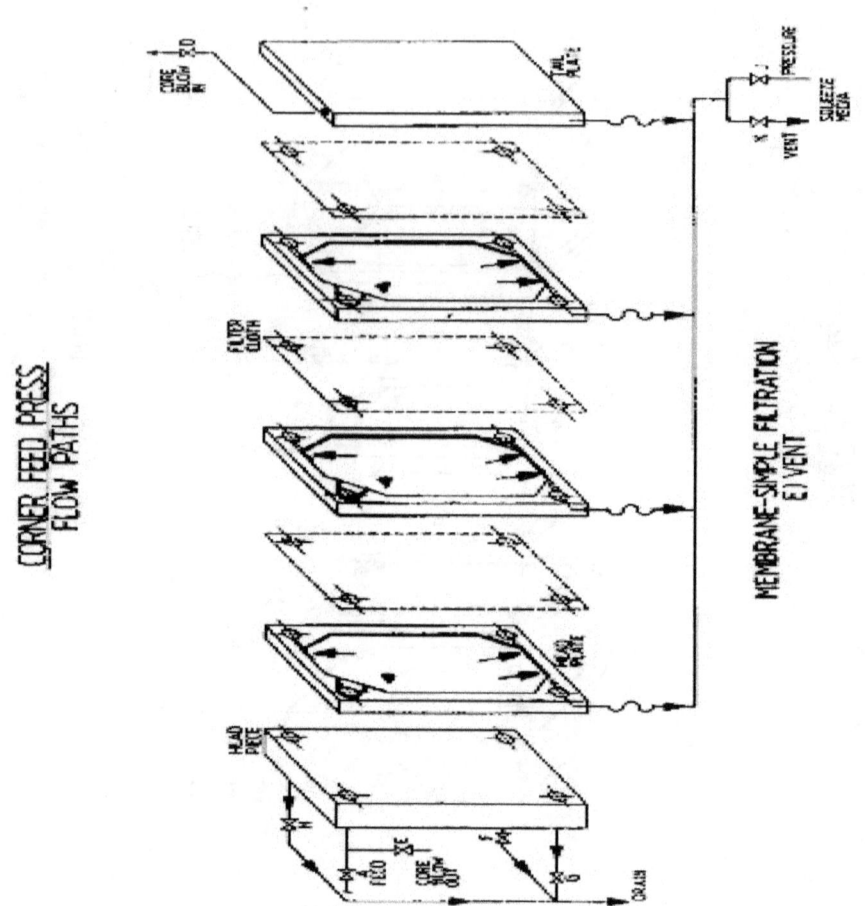

LIQUID SOLID SEPARATION WITH FILTER PRESSES

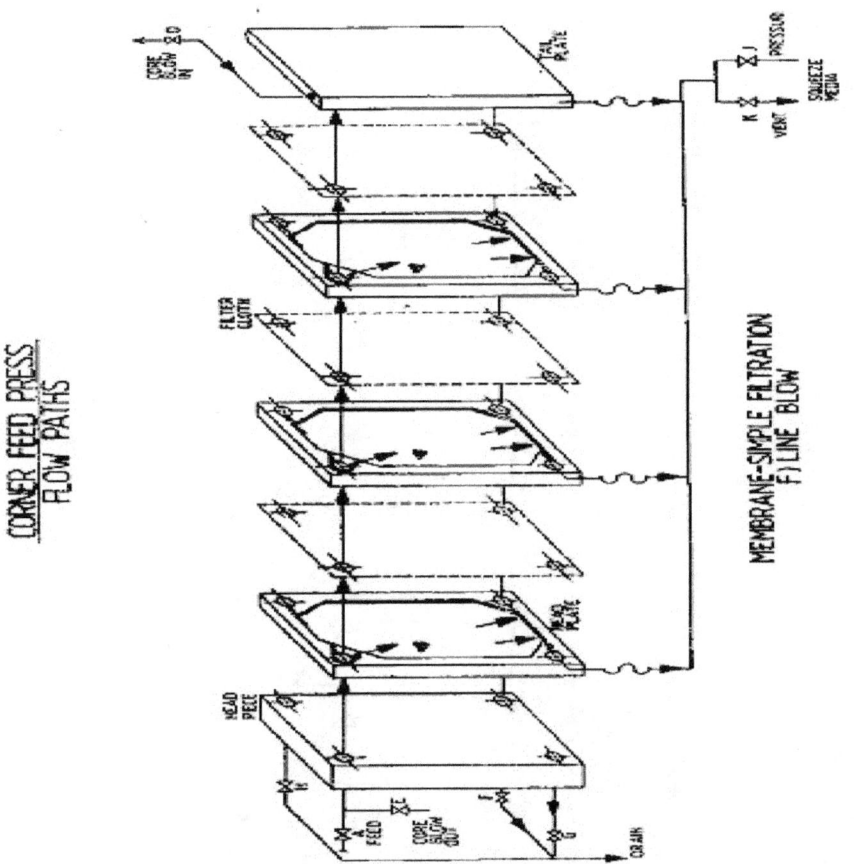

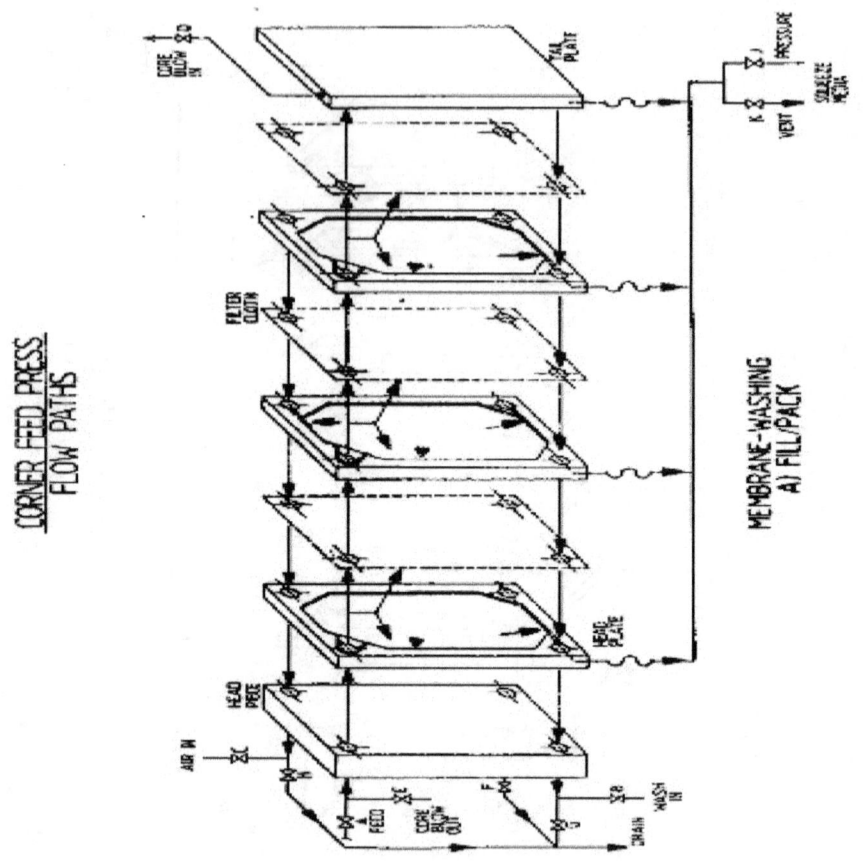

LIQUID SOLID SEPARATION WITH FILTER PRESSES

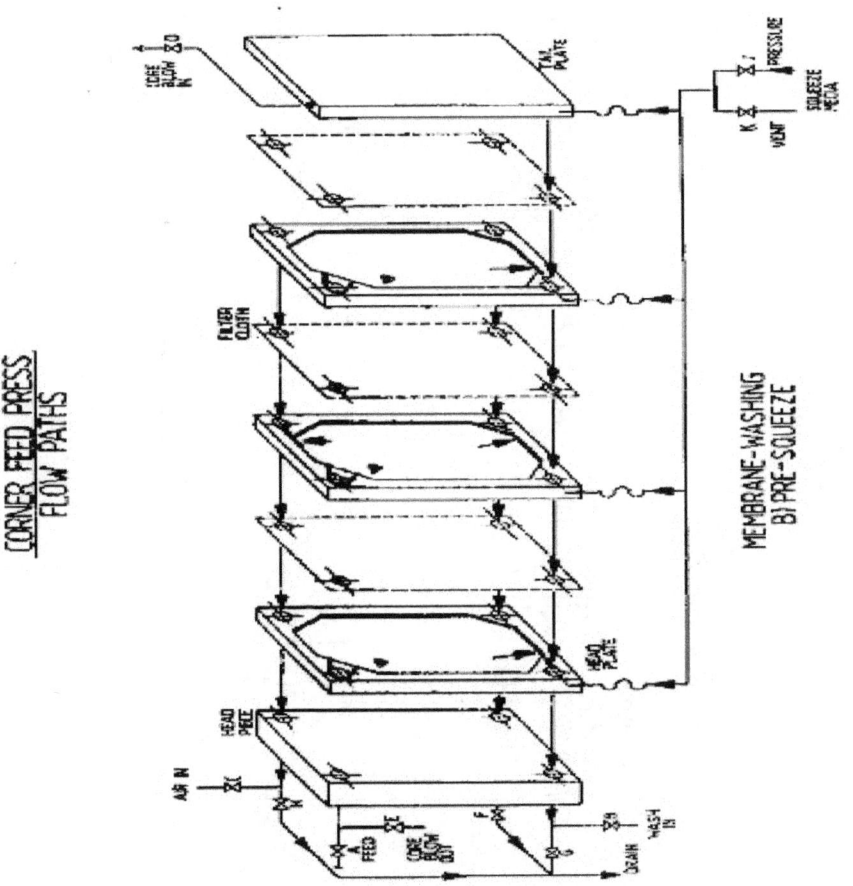

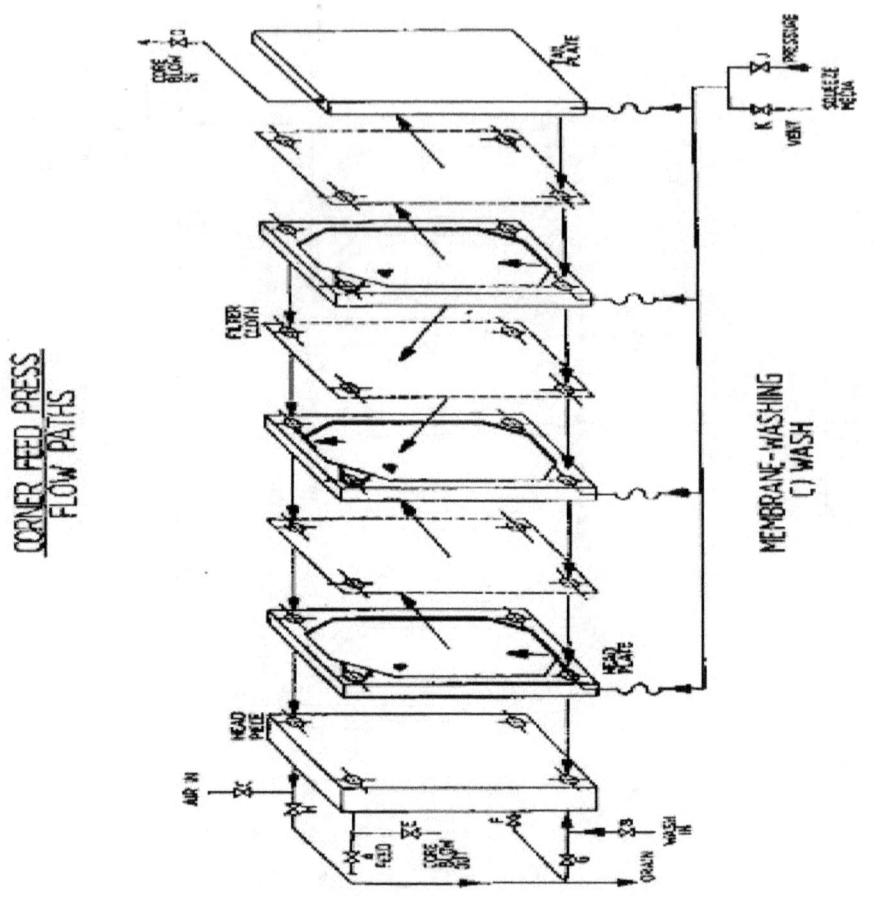

LIQUID SOLID SEPARATION WITH FILTER PRESSES

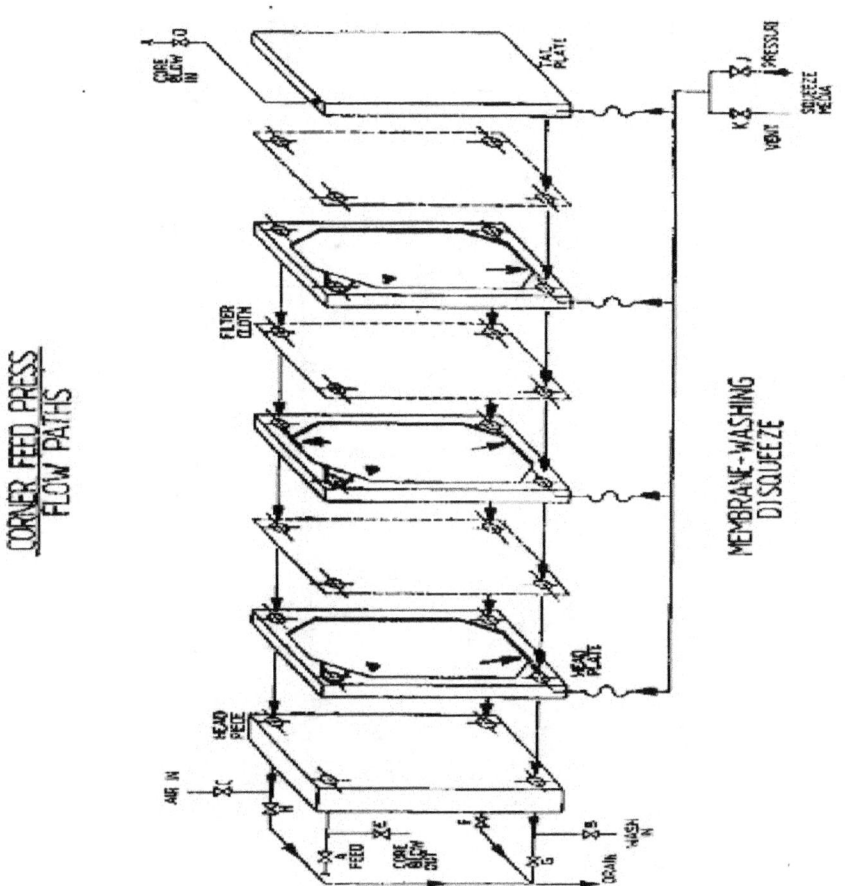

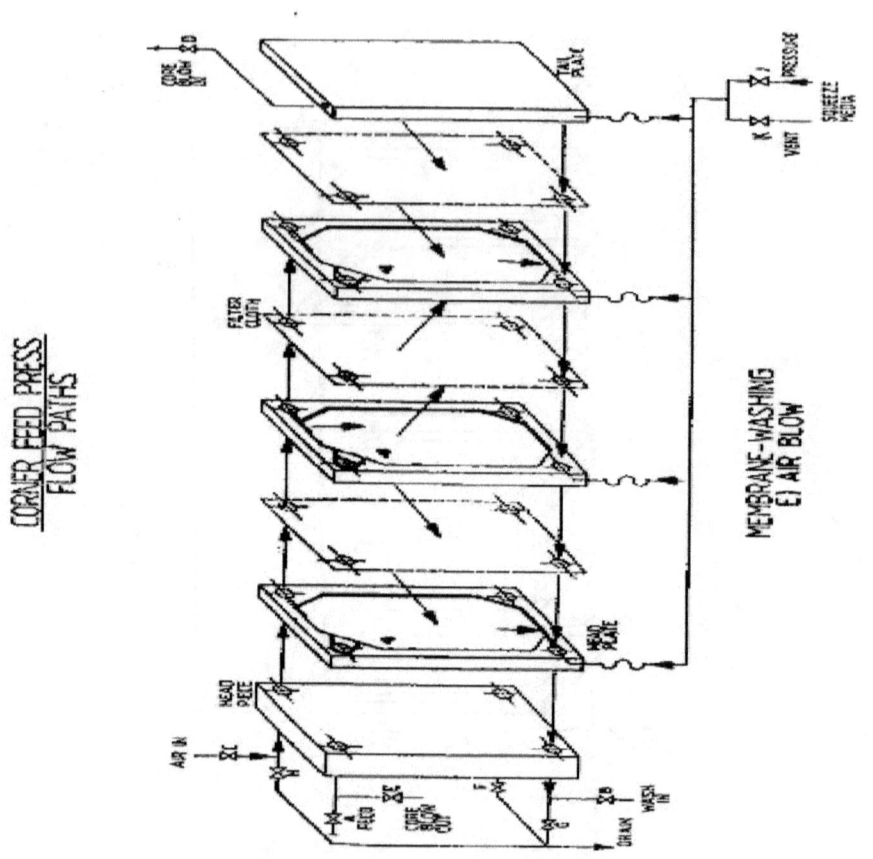

LIQUID SOLID SEPARATION WITH FILTER PRESSES

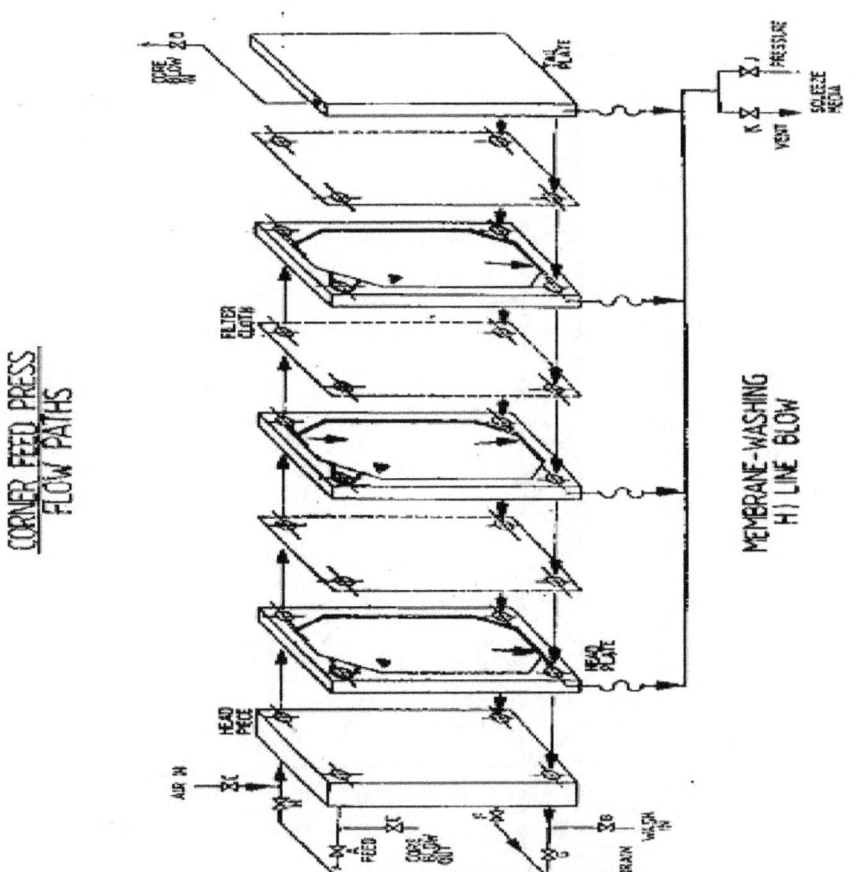

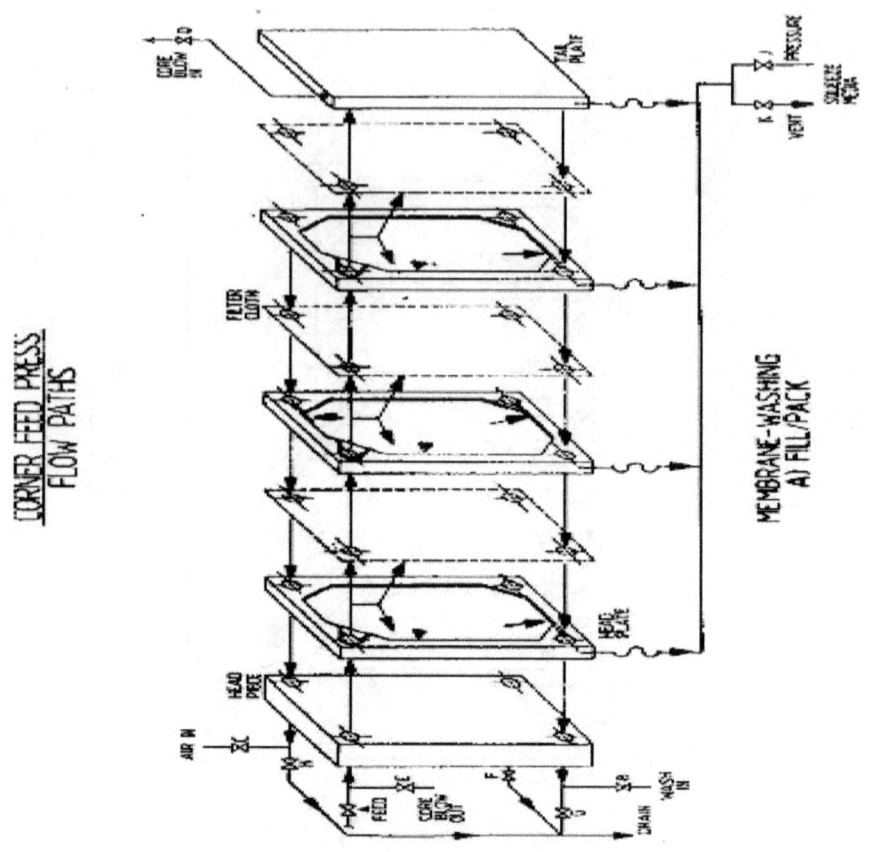

LIQUID SOLID SEPARATION WITH FILTER PRESSES

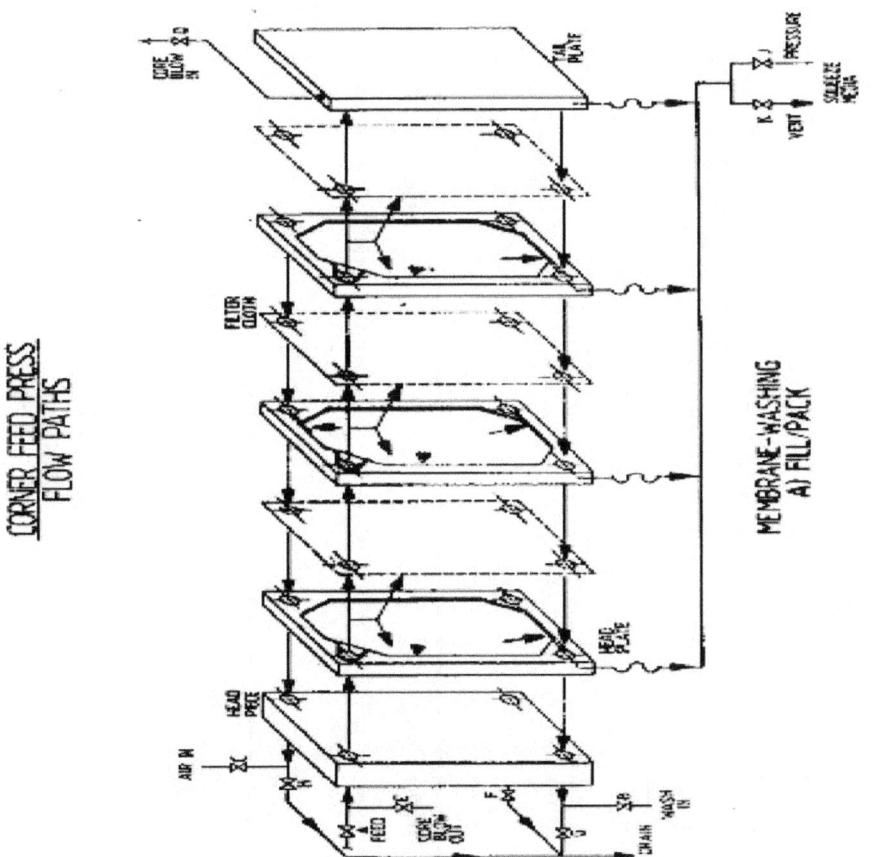

VALVE CHARTS - CENTRE FEED PRESSURE

SIMPLE FILTRATION - FIXED VOLUME

	A	B	C	D	E	F	G	H	I	J	K
FILL	0	-	-	-	-	X	X	0	0	-	-
PACK	0	-	-	-	-	0	0	0	0	-	-
PRE - SQUEEZE											
WASH											
SQUEEZE											
AIR BLOW											
CORE BLOW											
VENT											
LINE BLOW											

WASHING/AIR BLOWING - FIXED VOLUME

	A	B	C	D	E	F	G	H	I	J	K
FILL/PACK	0	X	X	-	-	X	X	0	0	-	-
	0	X	X	-	-	0	0	0	0	-	-
PRE - SQUEEZE											
WASH	X	0	X	-	-	X	X	0	X	-	-
SQUEEZE											
AIR BLOW	X	X	0	-	-	X	0	X	X	-	-
CORE BLOW											
VENT											
LINE BLOW	X	X	0	-	-	0	0	X	X	-	-

SIMPLE FILTRATION - FIXED VOLUME

	A	B	C	D	E	F	G	H	I	J	K
FILL	0	-	-	X	X	X	X	0	0	X	0
PACK	0	-	-	X	X	0	0	0	0	X	0
PRE - SQUEEZE											
WASH											
SQUEEZE	X	-	-	X	X	0	0	0	0	0	X
AIR BLOW											
CORE BLOW	X	-	-	0	0	X	X	X	X	0	X
VENT	X	-	-	X	X	0	0	0	0	X	0
LINE BLOW	X	-	-	0	X	X	X	X	X	X	0

WASHING/AIR BLOWING - FIXED VOLUME

	A	B	C	D	E	F	G	H	I	J	K
FILL/PACK	0	X	X	X	X	X	X	0	0	X	0
	0	X	X	X	X	0	0	0	0	X	0
PRE - SQUEEZE	X	X	X	X	X	X	0	0	0	0	X
WASH	X	0	X	X	X	X	X	0	X	0	X
SQUEEZE	X	X	X	X	X	0	0	0	0	0	X
AIR BLOW	X	X	X	0	X	X	0	X	X	0	X
CORE BLOW	X	X	X	0	0	X	X	X	X	0	X
VENT	X	X	X	X	X	0	0	0	0	X	0
LINE BLOW	X	X	0	X	X	0	0	X	X	X	0

0 = OPEN
X = CLOSED
- = NOT APPLICABLE

LIQUID SOLID SEPARATION WITH FILTER PRESSES

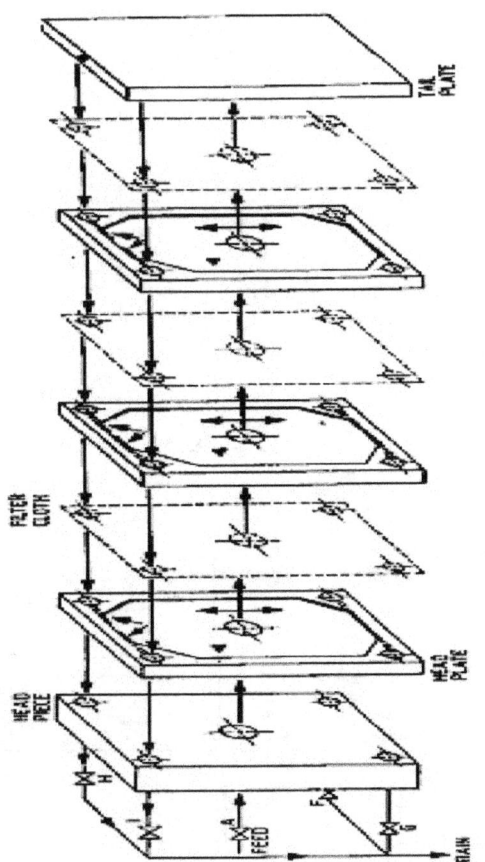

HANS G. VALERIUS PHD

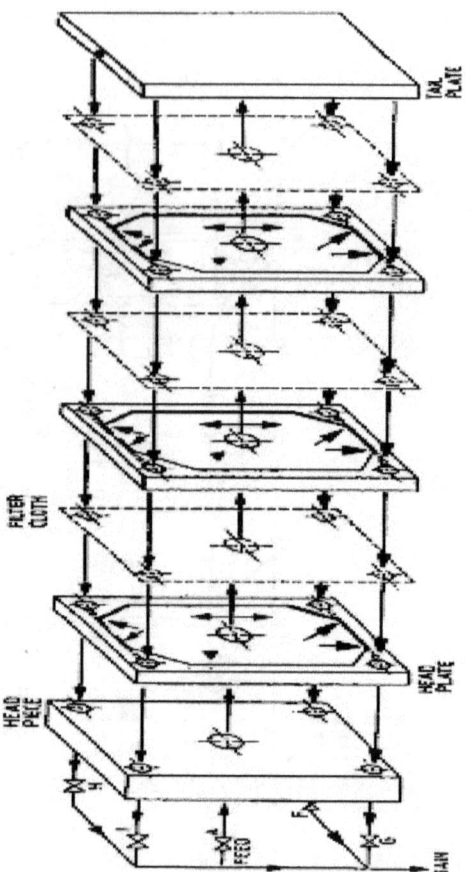

LIQUID SOLID SEPARATION WITH FILTER PRESSES

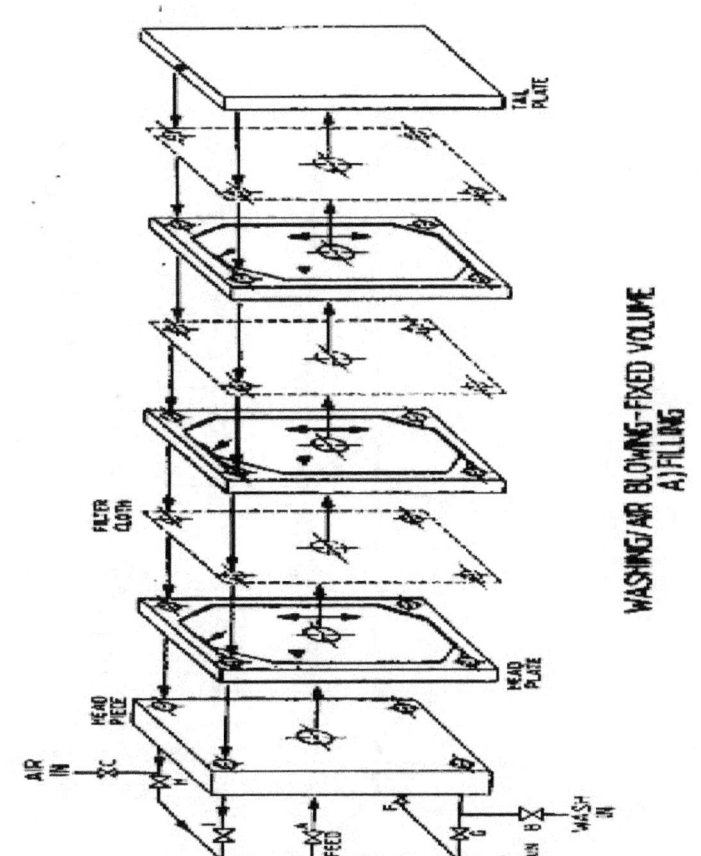

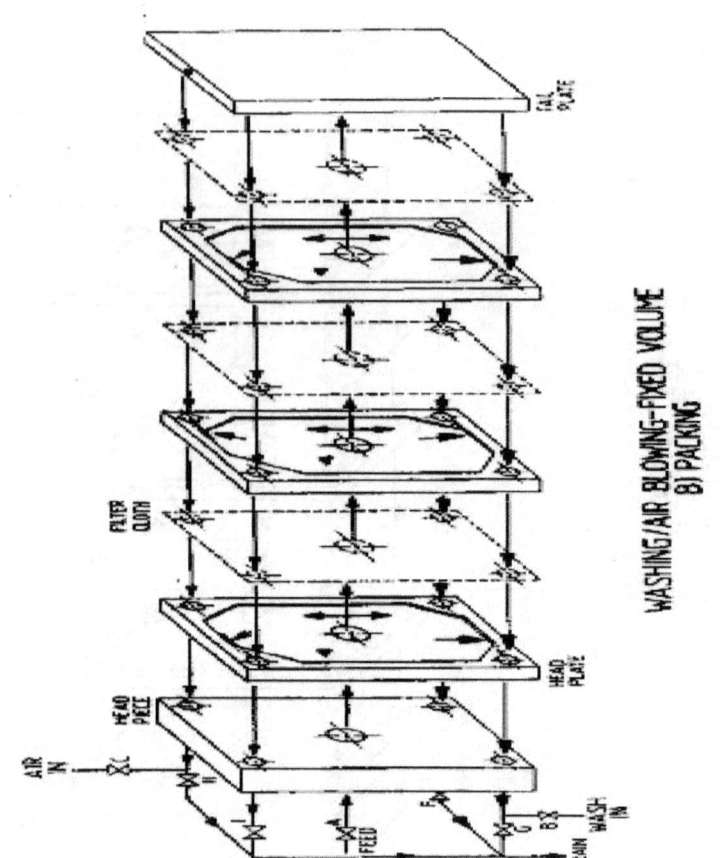

LIQUID SOLID SEPARATION WITH FILTER PRESSES

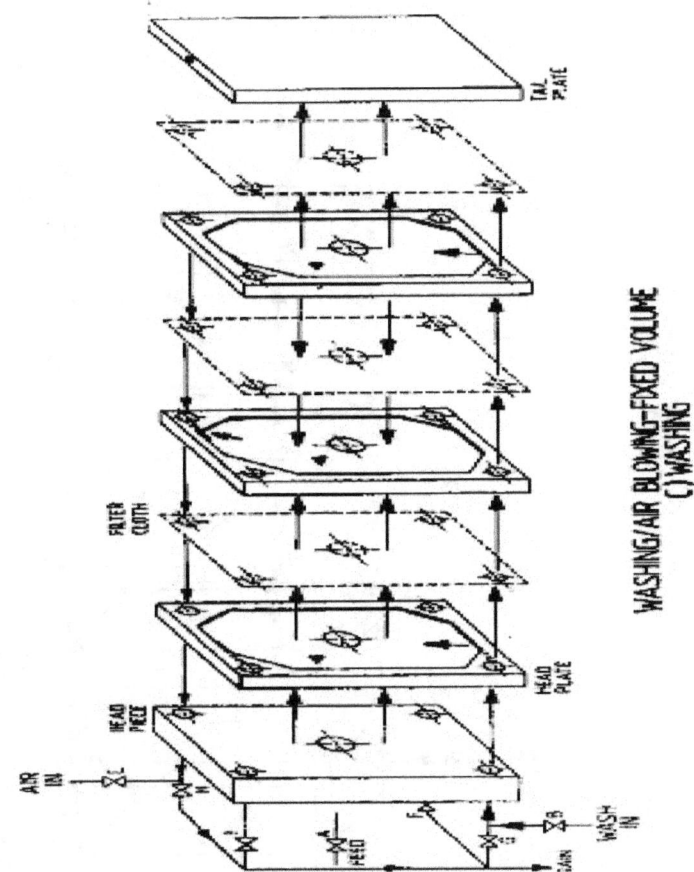

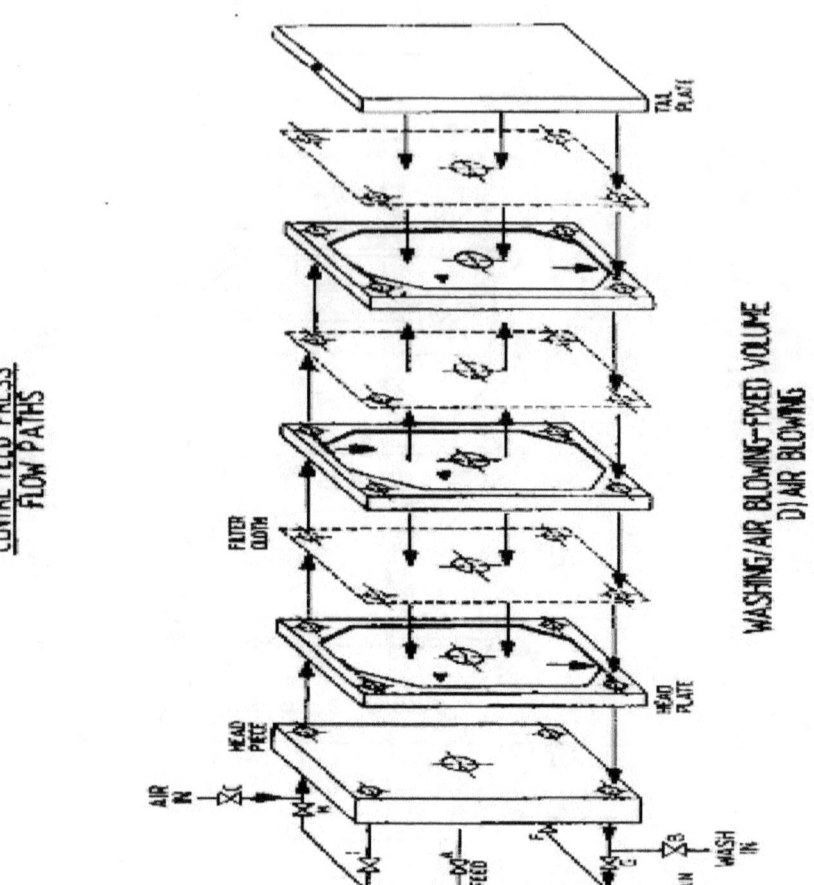

LIQUID SOLID SEPARATION WITH FILTER PRESSES

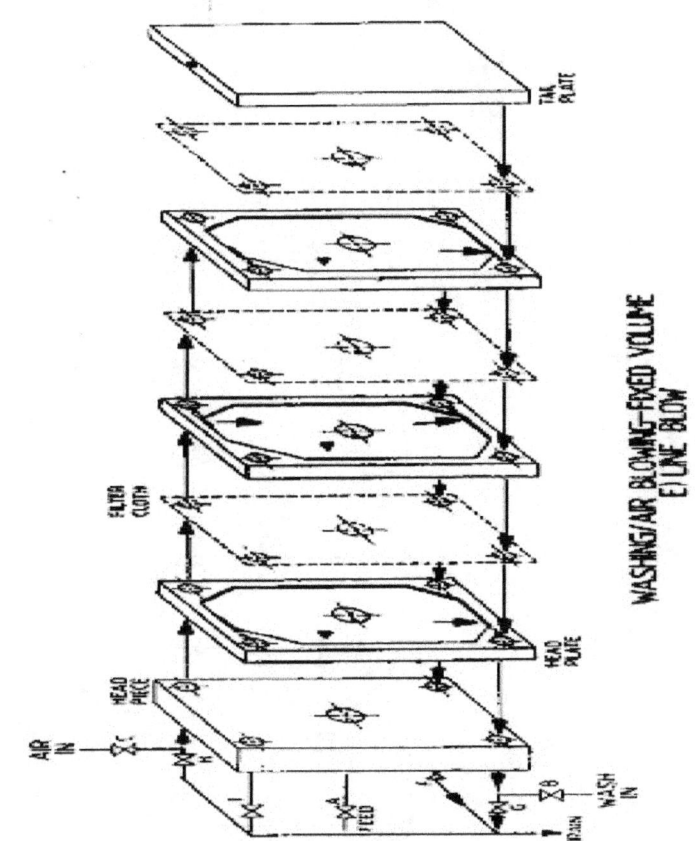

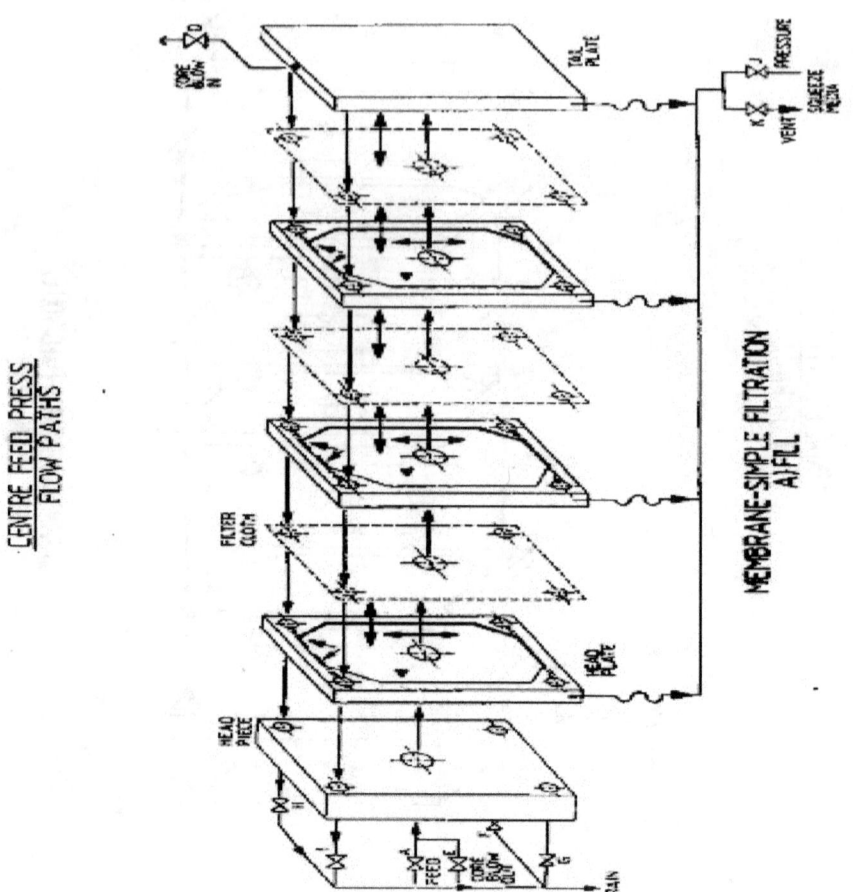

LIQUID SOLID SEPARATION WITH FILTER PRESSES

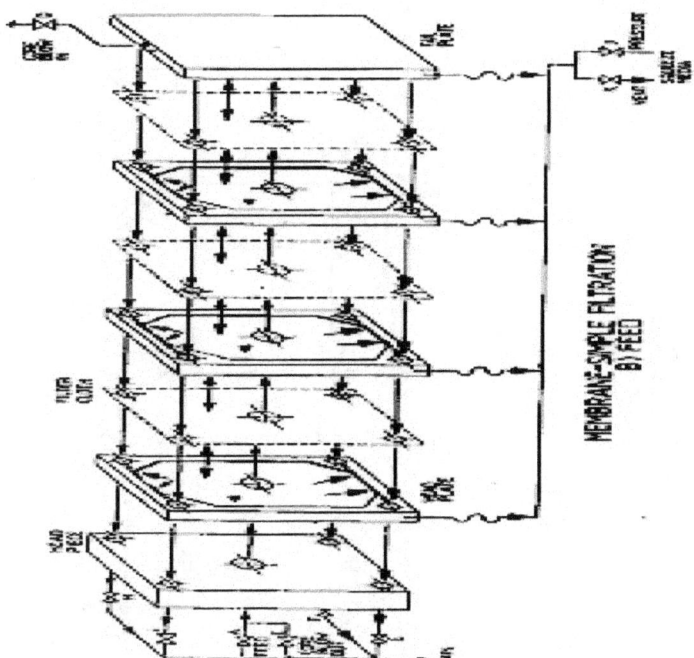

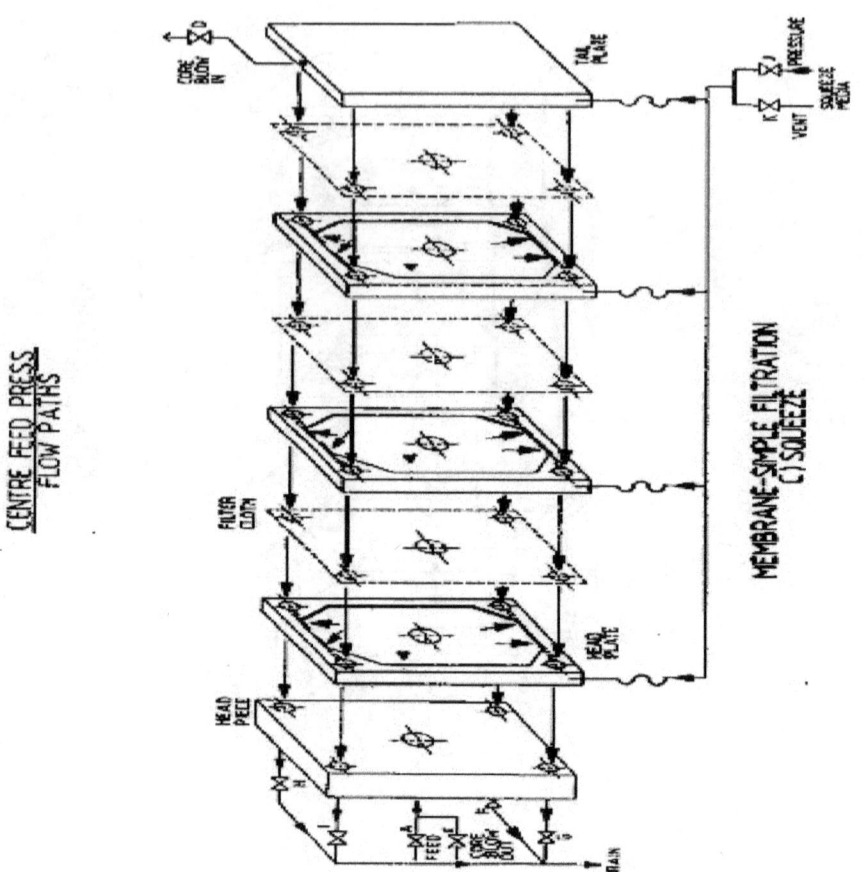

LIQUID SOLID SEPARATION WITH FILTER PRESSES

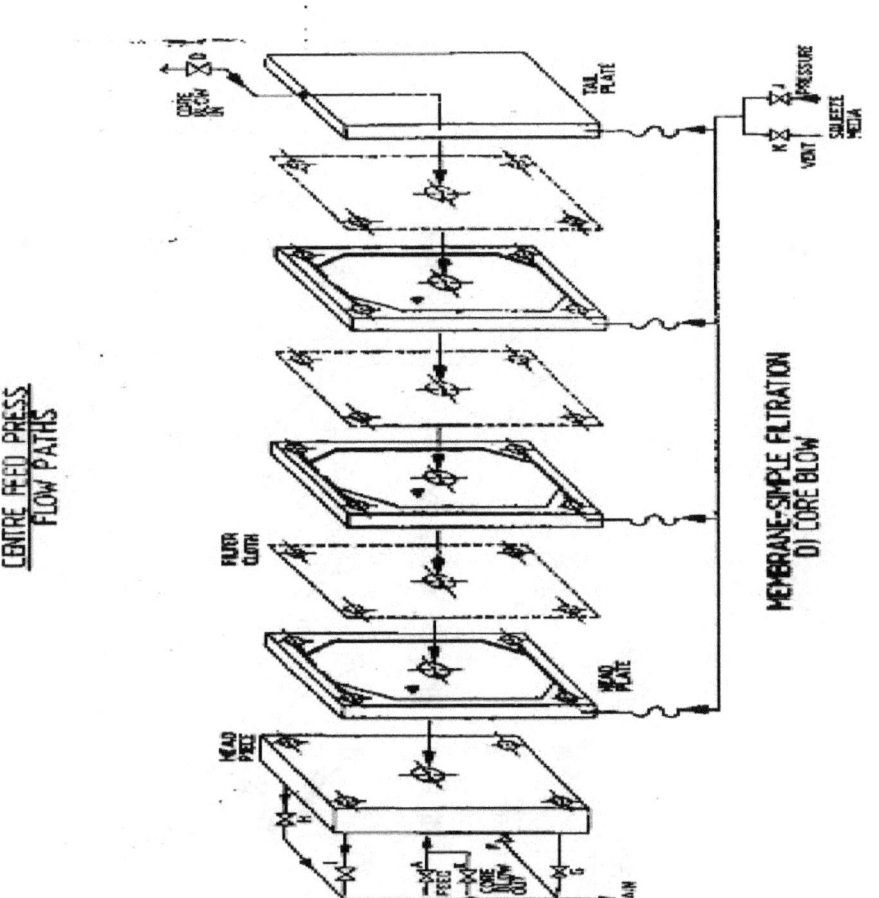

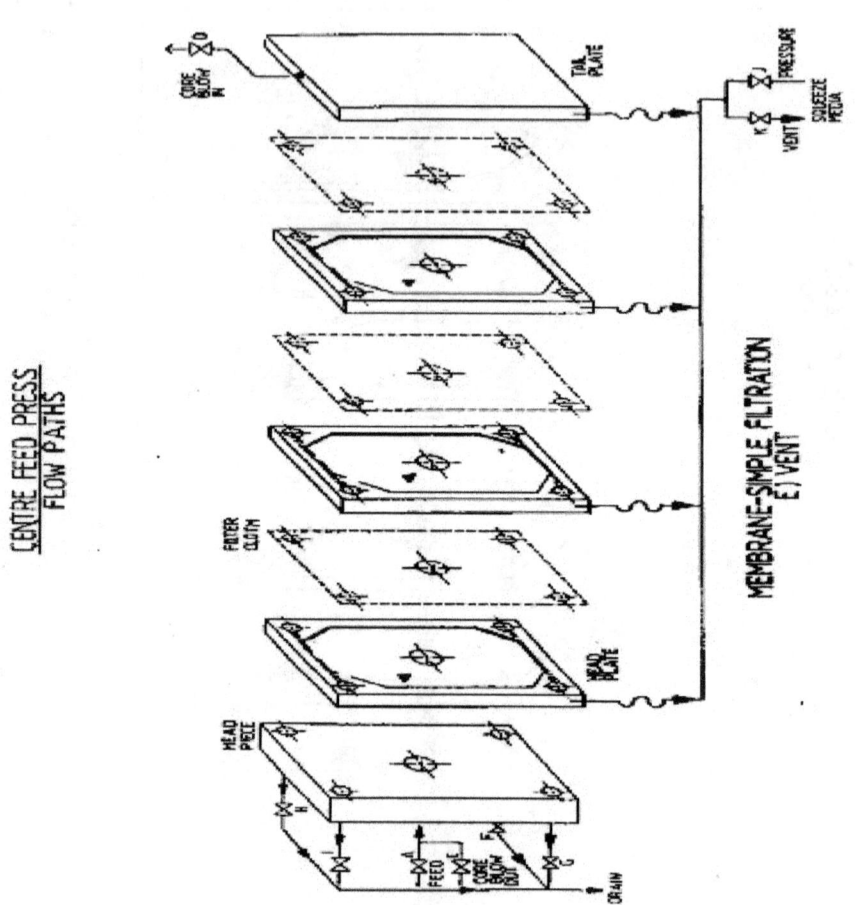

LIQUID SOLID SEPARATION WITH FILTER PRESSES

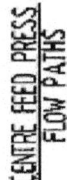

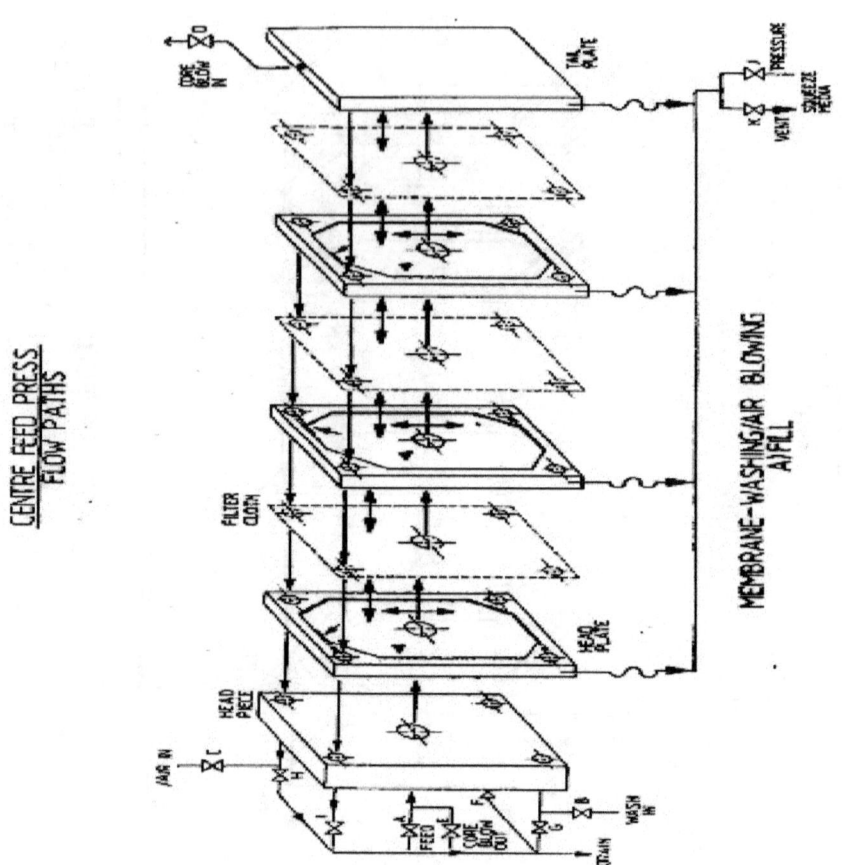

LIQUID SOLID SEPARATION WITH FILTER PRESSES

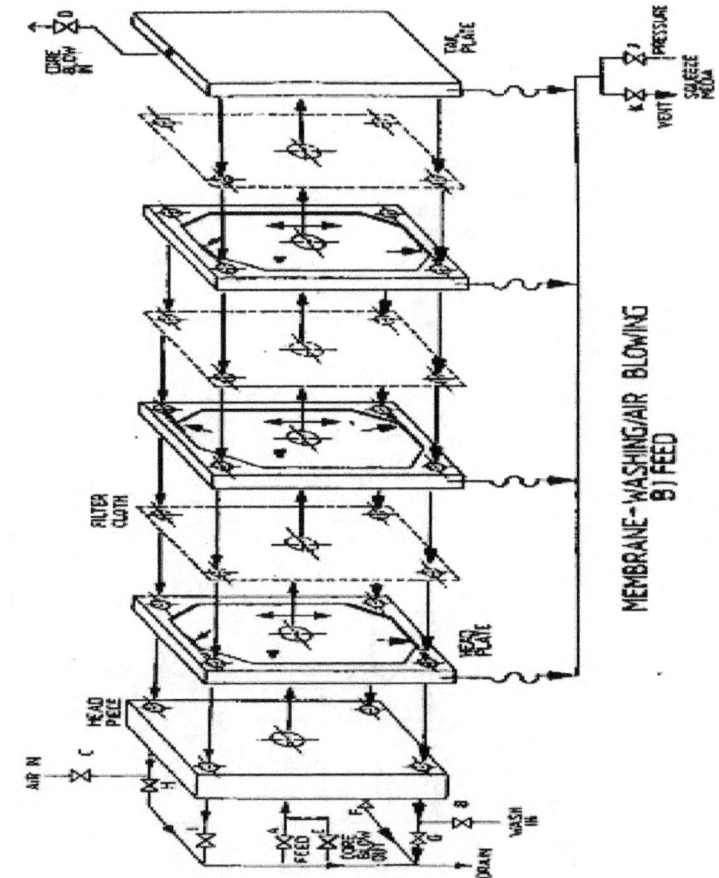

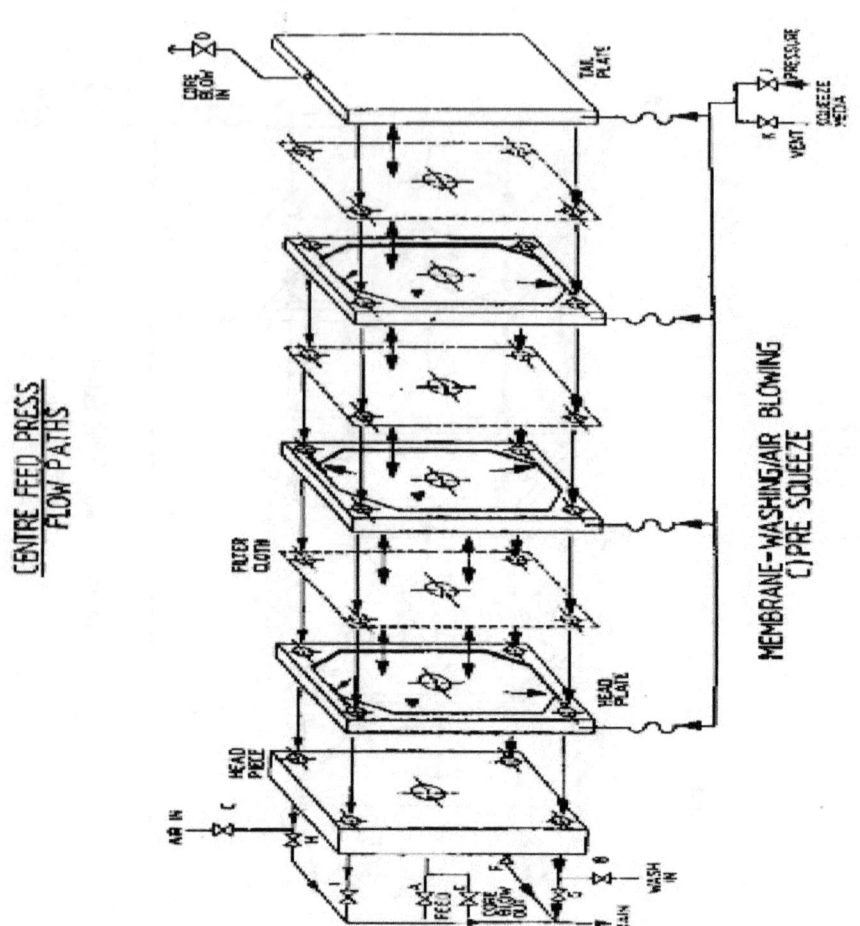

LIQUID SOLID SEPARATION WITH FILTER PRESSES

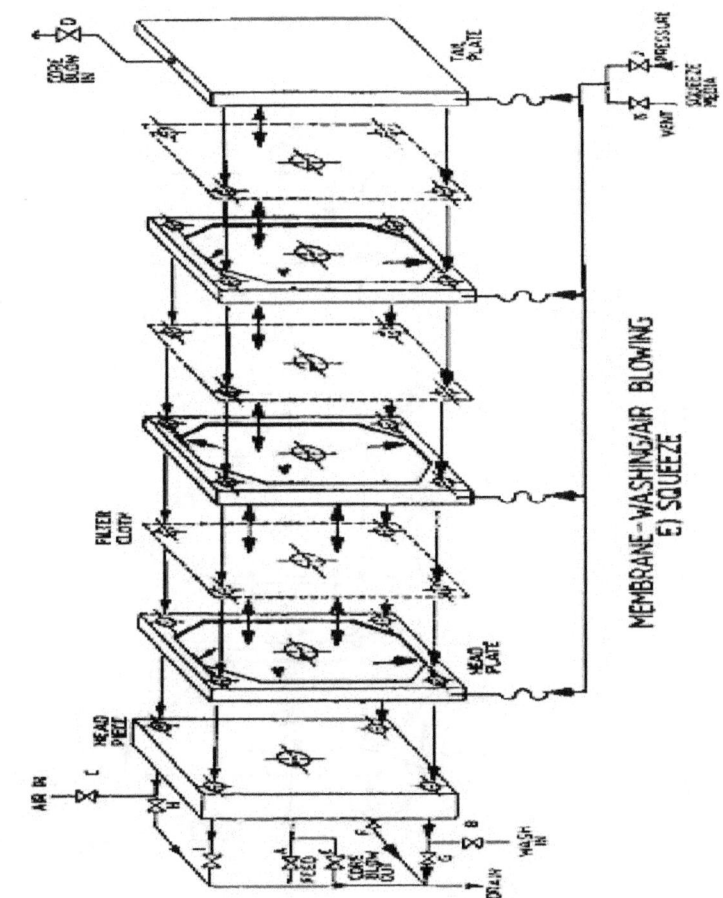

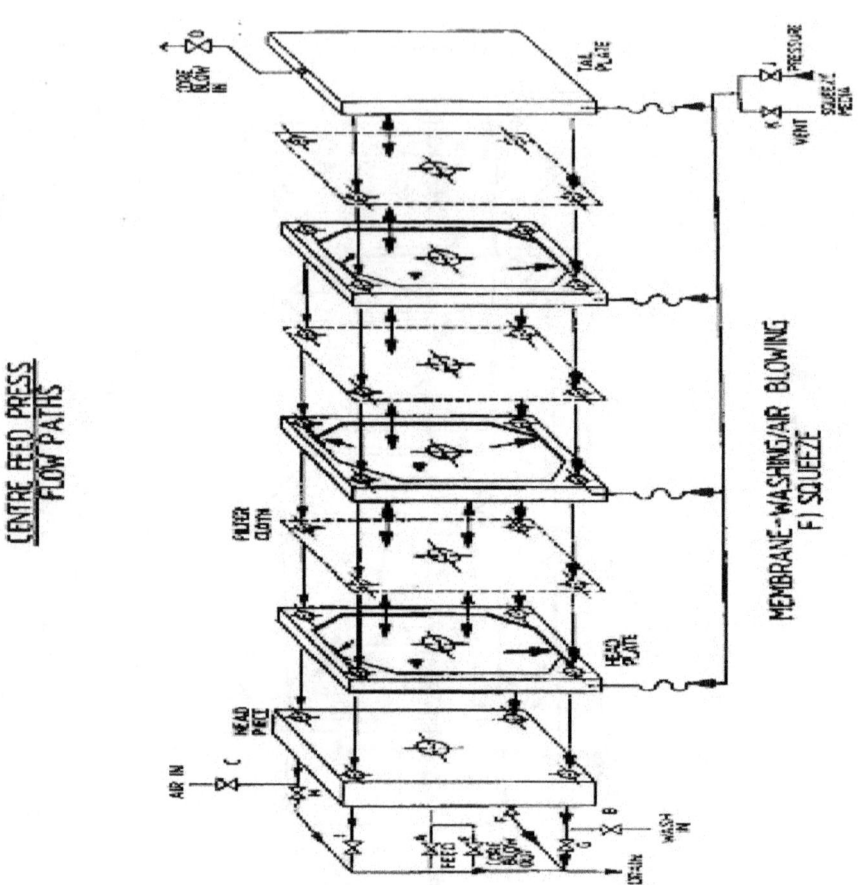

LIQUID SOLID SEPARATION WITH FILTER PRESSES

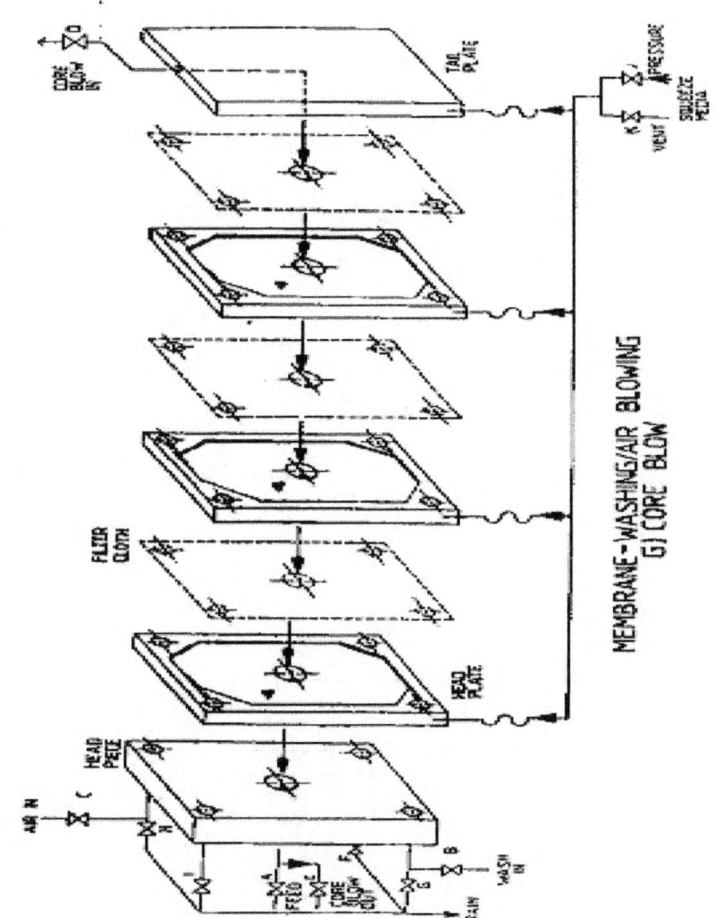

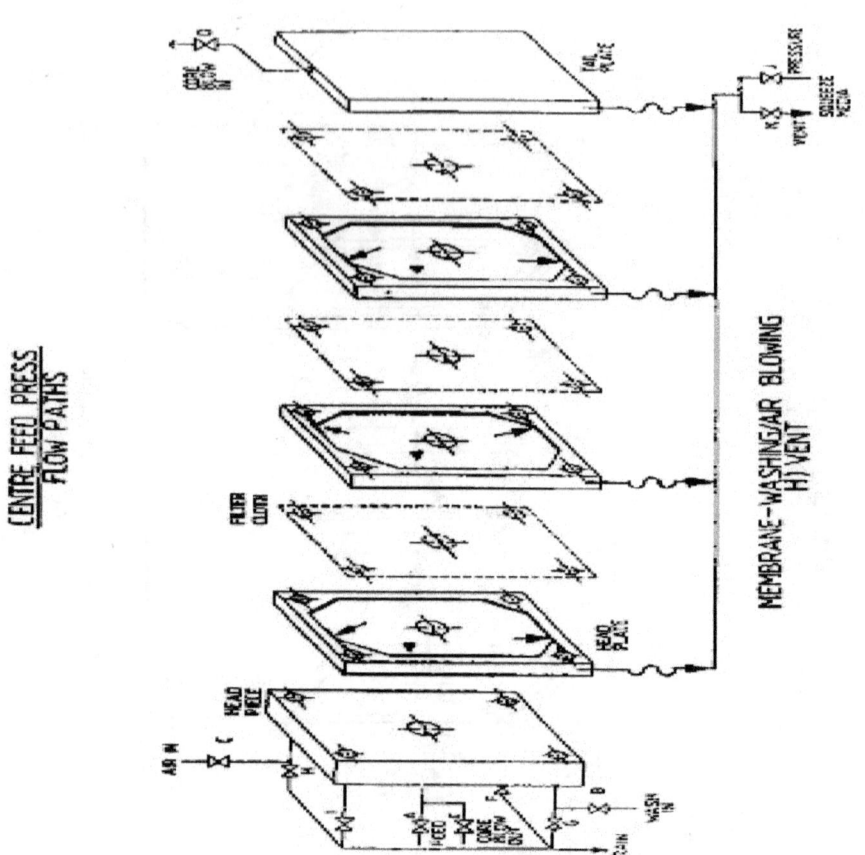

LIQUID SOLID SEPARATION WITH FILTER PRESSES

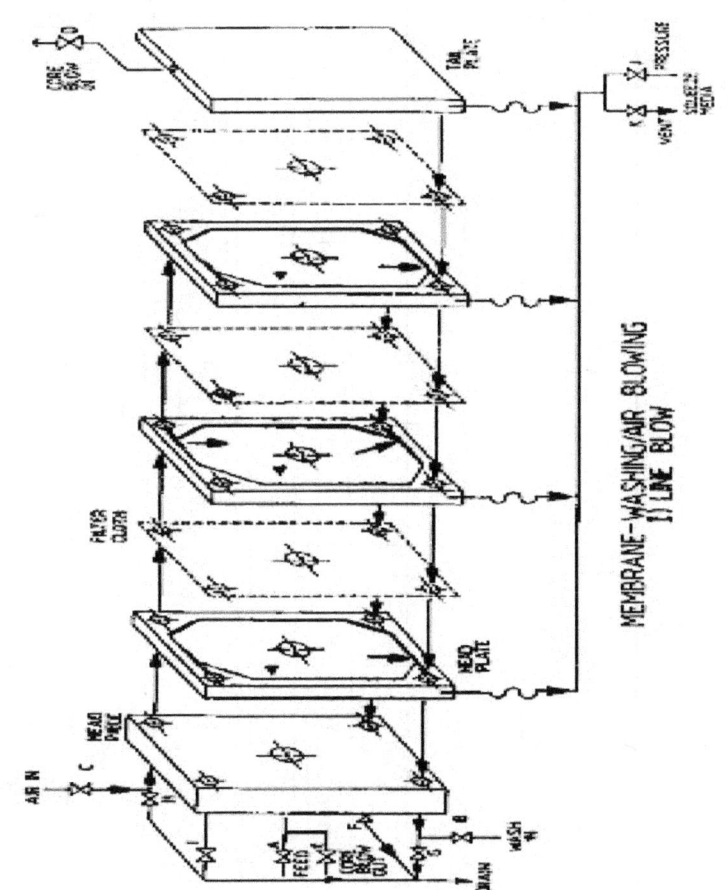

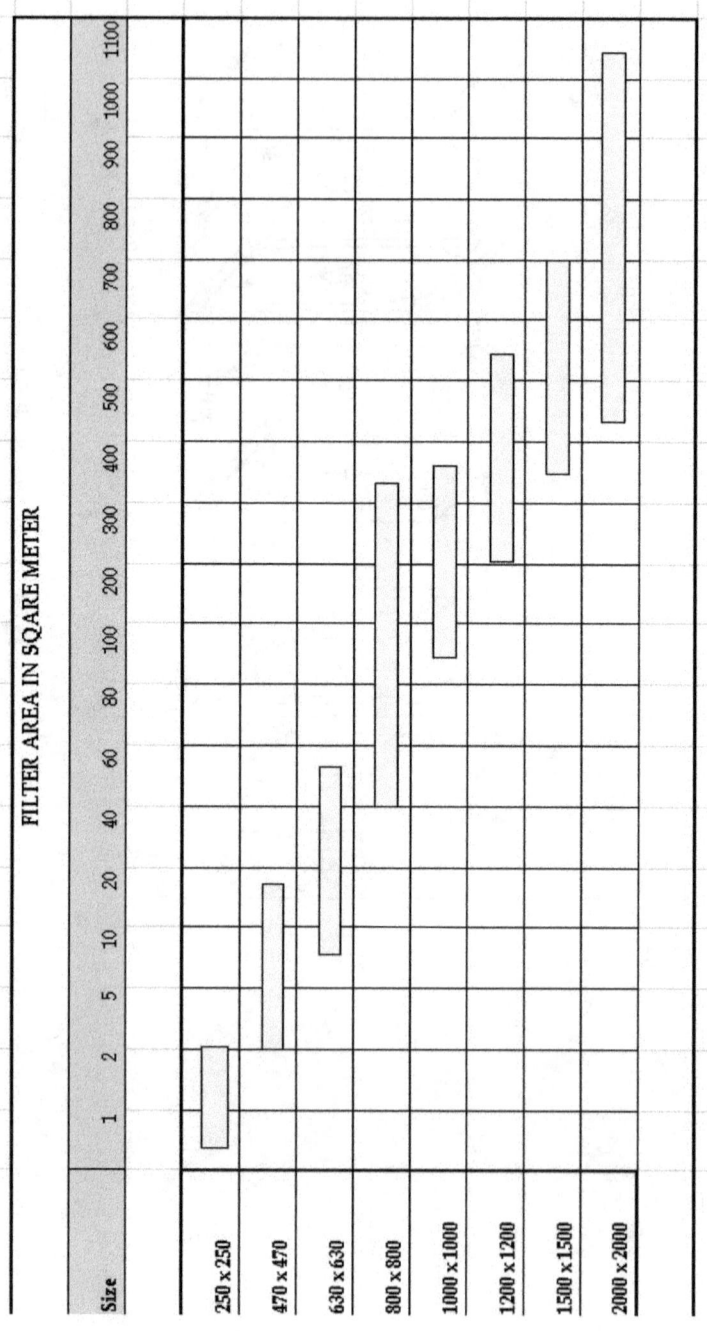

LIQUID SOLID SEPARATION WITH FILTER PRESSES

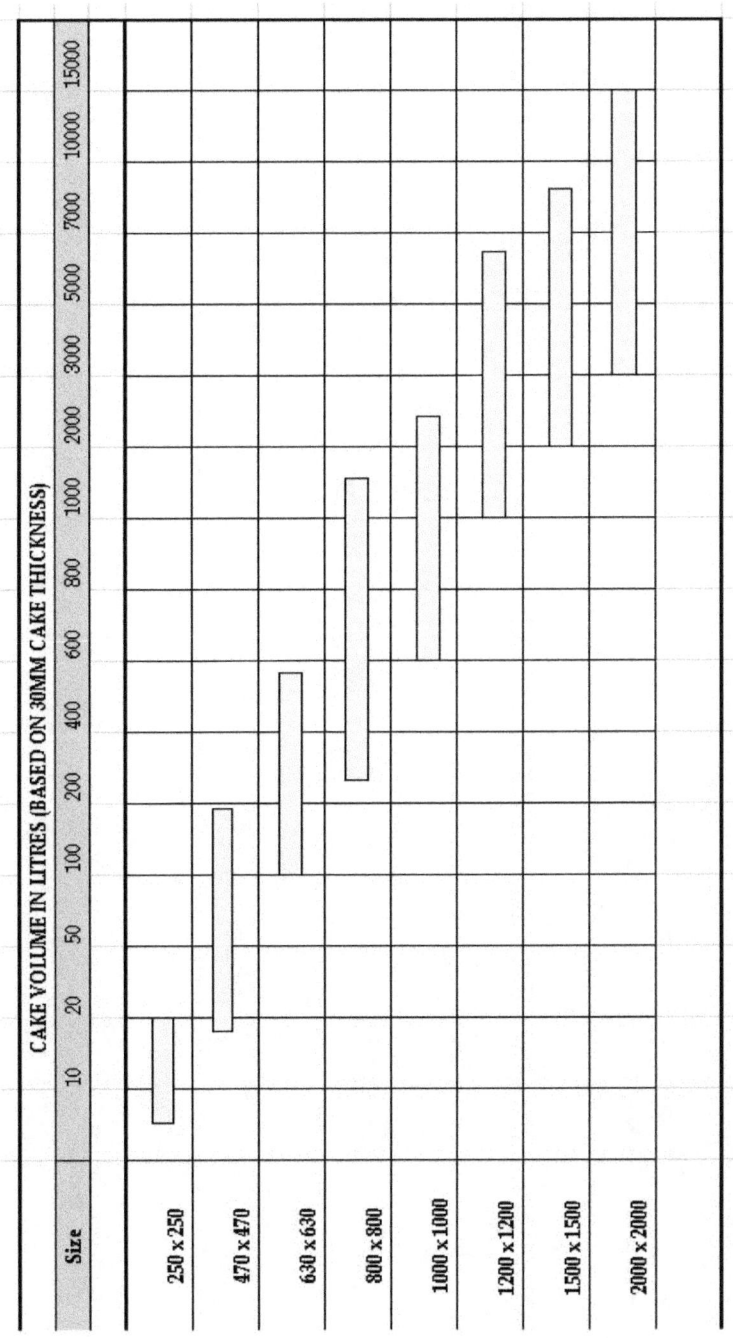

www.ingramcontent.com/pod-product-compliance
Lightning Source LLC
Chambersburg PA
CBHW071421170526
45165CB00001B/350